HOUSE....... ....No. 13.

REPORT OF COMMISSIONERS

CONCERNING

AN AGRICULTURAL SCHOOL.

JANUARY, 1851.

Commonwealth of Massachusetts.

The Commissioners to whom were referred the Resolves of the last Legislature, concerning the establishment of an Agricultural School, and other subjects relative to the advancement of the interests of agriculture in this Commonwealth, submit the following

REPORT:

The first seed ever planted was the first effort of civilization. The highest physical triumph of civilization, therefore, must be, that perfection in the art of cultivation which will afford abundant sustenance to the largest possible population.

Agriculture is that art; the nursing mother of nations. With its prosperity, population multiplies, commerce and manufactures increase, all the industrial pursuits of mankind flourish, and wealth and comfort abound. On the contrary, let the cultivation of the soil in any nation be neglected, the hum of business will be silenced, the arm of industry paralyzed, and both individual and national happiness destroyed. Or, let the earth cease to yield her annual increase; yea, let but one of her accustomed crops be cut off, as we have witnessed in Ireland and other countries, and how soon scenes of want, misery and crime ensue, constraining multitudes to abandon home and country in search of sustenance in a foreign land, or consigning them, by famine and pestilence, to untimely graves.

The wisdom of WASHINGTON, the Father of his Country, was in nothing more conspicuous than in this remark: "I know of no pursuit in which more real or important good can be rendered to a country, than by the improvement of its agriculture;" a sentiment which we rejoice to see reiterated in one of the late public documents from the capital bearing his venerable name.

"The agricultural interest stands first in importance in our country, and embodies within itself the principal elements of our national wealth and power; and it should be with us, as it has been, and is, with all other prosperous civilized nations, a leading object of public regard."

This art is, indeed, the primitive and most important pursuit of man. On its success depends the welfare, not only of one nation, but of the whole civilized world. Its importance can never be appreciated until we arrive at the final results of commerce, and the other great industrial pursuits which rest upon it, nor until we can obtain the aggregate of those blessings which it has conferred, and is capable of conferring, on the human race.

It should, therefore, be the especial care of every government to promote among its subjects such a knowledge of this world-sustaining art, as to enable them to derive their support from its soil, and thus to make them an independent as well as a happy people. Upon this principle our fathers acted, and the wisdom of their legislation sustains this remark.

Besides, increasing attention to this art is demanded by the character of the present age, so distinguished for scientific discoveries, and the facility with which these are applied to the other arts of life. For instance, what wonderful improvements have been witnessed in the art of printing, since the apprenticeship of our Franklin; in the manufacture of cloth, since the inventions of Arkwright and others; in the comfort and velocity of travelling, since the discoveries of Fulton; and in international communication, by the lightning speed of telegraphic communication;—*all*, ALL by the application of science to art.

And has she no contribution to make in aid of agriculture, the parent of all arts? True, she has already done something, in the improvement of labor-saving implements, to lessen the toil of the farmer, and enable him to perform in an hour the work of a day; and she needs but a wise direction of thought, enterprise, and capital, to accomplish still greater results.

The triumphs of inventive genius have placed, on the bright page of American history, names illustrious for patriotism, science, and art. There are many whom we honor and revere as pioneers and practical men in agriculture; but, in the application of profound scientific knowledge to this time-honored art, where are the men who are to stand on the roll of fame by the side of Rittenhouse, Franklin, and Fulton? If "knowledge is power," the want of it is weakness; and this axiom is as applicable to agriculture as to any other employment. The doors of the temple of science are open to all who desire to enter it, and there is no reason why the farmer, as well as the mechanic or manufacturer, may not pass its portals.

The investigations of scientific men have proved, beyond the possibility of a doubt, that, by the analysis of the soil, and the desired crop, and a wise reference to atmospheric influences, we are as competent to adapt food to the different species of vegetables, as to the various kinds of animals; for instance, to feed a crop of corn, as a herd of swine.

An example will illustrate this remark. In a letter from Professor Mapes, the scientific editor of the "Working Farmer," addressed to General Tallmadge, President of the American Institute, is the following statement:—

"During the last winter, I made an analysis of soil from a field which refused corn last year, and found the soil deficient in the following constituents: chlorine, soda, phosphoric acid, lime, potash, and ammonia. The last spring, I applied a compost of common salt, decomposed by lime, thus supplying chlorine and soda; spent bone dust, of the sugar refiners, which furnished phosphoric acid; Peruvian guano, containing potash and ammonia, to which was added a small portion of charcoal dust and plaster of Paris, to retain the volatile portions.

"The above was added to the soil at an expense of *one dollar and thirty-one cents per acre*, and the field planted with corn. The crop is now standing, and the Committee of the American Institute on Farms will state to you that the crop will probably be from fifty to seventy-five bushels of shelled corn per acre."

Professor Mapes further states, that in no instance has the

experiment failed to produce desired crops, of superior quality, where manuring has been founded on the chemical constituents of the soil, &c. Among these he mentions several instances where corn has produced over one hundred bushels per acre; wheat, forty to fifty-seven bushels per acre; potatoes, three hundred to four hundred bushels per acre; carrots, one thousand bushels; ruta baga, twelve hundred bushels; and other crops in proportion.

Similar facts have been developed by the mere rotation of crops, instances in which lands had produced abundantly, without the application of manure, for several years.

This theory teaches, that certain products are adapted to certain soils, and that, where particular ingredients have been exhausted from the soil by vegetation, the indiscriminate use of fertilizing materials will not necessarily ensure a crop.

Already the exhaustive process of perpetual cropping has travelled over the once fertile lands of New England, and in its desolating march is wending its way over the fair fields of New York, Ohio, and on to the "Far West." Under the influence of this system of cultivation, the crops of wheat in these States have receded from an average of twenty-two bushels to fourteen bushels, or less, per acre; and the same remark will apply to other crops, in like ratio of reduction.

From this sad but common error, Europe is just recovering; and, under the influence of her agricultural schools, now scattered all over the continent, (as will be seen by this report,) and of scientific cultivation, her crop of wheat in many parts has advanced from sixteen bushels to an average of over thirty bushels per acre; and a similar increase has taken place in other crops. Wonders have also been achieved in reclaiming waste lands, and in converting those which were barren and worthless, into rich and productive farms.

Our farmers are becoming aware of these facts, but do they realize that the present system of impoverishing our lands, without sustaining their natural strength and fertility, will, sooner or later, end in barrenness? and that, if the present population may rightfully exhaust one-third part of the arable lands of the United States of their natural fertility, the population which will be here at the close of the present century will, long before that period, have consumed the remaining two-thirds of all American territory. By a calculation which appears in a late Report of the Patent Office at Washington, it is estimated that *one thousand millions of dollars* would not more than restore to their original richness and strength the *one hundred million acres of lands* in the United States, which have already been partially exhausted of their fertility.

One-half of the present century has elapsed, and in it, nearly two generations have passed off the stage; a period, too, distinguished for scientific discoveries, and for the progress of civilization; yet, in this boasted age of improvement, no institution exclusively for instruction in agriculture has been established, either by this Commonwealth, or by any other State of the Union. But dark as this view of agricultural education really is, says Dr. Lee, in his Report to the Secretary of the Interior, it is but the dawn of a bright and happy day. "Many who have labored for the improvement of agriculture and the education of the agriculturists, for a quarter of a century, with little hope of reward, now realize the beginning of an auspicious change in public sentiment. Thanks to agricultural societies and journals, the people will soon discover that labor and capital devoted to tillage and husbandry, are as worthy of legislative consideration and aid, as when applied to mining, commerce, and manufactures.

"It is indeed wonderful how long those enlightened farmers, who, like Washington, cherish a due respect for their high calling, have had to beg, and beg in vain, of State Legislatures and of Congress, for a little assistance to prevent the universal impoverishment of American soils. Neither the earnest recommendation of the illustrious farmer of Mount Vernon, nor the prayers of two generations of agriculturists, nor the painful fact, that nearly all the tilled lands were becoming less and less productive, could induce any legislature to foster the study of agriculture as a *science*. Happily, this term, when used in connection with rural affairs, is no longer the subject of ridicule.

"A great principle is involved in the science of agriculture, which reaches through indefinite generations, and forms the basis of all possible improvements and of the highest hopes of our race. As a nation of farmers, is it not time that we inquire by what *means* and on what terms the fruitfulness of the earth, and its invaluable products may be forever maintained, if not forever improved.

" These are questions of universal concernment, to the careful and rigid investigation of which, no man should refuse to lend a listening ear. A governmental policy which results in impoverishing the natural fertility of the land, must have an end. It is only a question of time, when this truly spendthrift course, this abuse of the goodness of Providence, shall meet with its inevitable punishment.

" A lack of mental culture and discipline is the most serious impediment to the diffusion of agricultural science among the farmers. Its language to them is an unknown tongue. Hence, the most sublime truths in the economy of nature are shut out from the popular understanding. It is feared that this will ever be the case, until schools designed to teach these branches of learning, which the practical farmer greatly needs but does not possess, are established and maintained throughout the United States."

In view of these remarks, your commissioners are of opinion that something should be done to arrest this evil, and to restore the great national loss occasioned by the misapplication of labor and capital in farming ; and that the time has arrived, when public sentiment also demands that some measures should be adopted by our Legislatures, as well as by Congress, to afford scientific and practical instruction in the various departments of agricultural husbandry.

Colleges and schools for this purpose have been established in Europe, sharing largely in the patronage, or wholly sustained by government, and under the immediate direction of learned and practical men. These may not be copies for our exact imitation, but models from which may be derived many valuable suggestions.

Impressed with the importance of laying before the Legislature all the information acquired in relation to these institutions, their systems of instruction and government, and the operation of the same, the commissioners submit the following communication as a part of this report, being the results of the investigations of a member of this Board, during his late tour in Europe. This valuable and able paper, the chairman may be permitted to say, embraces an amount of information and research never before presented to the American public, on this subject, and cannot fail to be read with great interest by all the friends of agricultural education:—

To the HON. MARSHALL P. WILDER, *Chairman of the Commissioners appointed by the Governor and Council of Massachusetts, on the subject of Agricultural Schools:*

SIR,—It was in the latter part of June, (1850) while in London, that I received from you a notice of my appointment as one of the Commissioners to consider the expediency of establishing an Agricultural School or Schools, in Massachusetts; and also a request that I would visit as many of the Agricultural Institutions in Europe as would enable me to understand the system of instruction, and the operation of these schools in that part of the globe. The notice reached me as I was about leaving England for the continent. But having accepted the appointment, I entered at once upon the examination suggested; first in England, then successively in Ireland, Scotland, Germany, Switzerland and France. The results of these investigations I now offer to the Commissioners, through you, as Chairman, rather than to the Executive, or Legislative departments of the government; because they were made at the request of the former, not of the latter.

Not anticipating any such request when I left home, I found myself not a little embarrassed in ascertaining the number, character and location of Agricultural Schools, so as to select the most important to be visited; for most of these schools being of recent origin, have not forcibly arrested public attention, like the Universities and Colleges of Europe. Nevertheless, through the kindness of gentlemen connected with the United States Legations abroad, and of some distinguished scientific friends in London, the subject gradually opened upon me fully, and I became more and more surprised to find how widely it now engages the attention of the governments, and of distinguished individuals, not only in the countries above named, but in several others in Europe, of which, also, I shall be able to give an account. I found, also, that when I had announced myself as from Massachusetts, making enquiries on the subject of education, I hardly needed any other introduction, so sensible are gentlemen abroad of the superiority of her

2

common school system; not hesitating, in many cases, to pronounce it the best on the globe.

Besides several gentlemen, who are at the head of the Agricultural Schools, and whose names will be mentioned when I come to describe those schools, it gives me pleasure to mention the following as having aided me very essentially in these investigations, and to whom, therefore, I am deeply indebted.

Hon. Abbott Lawrence, American Minister in London; Sir Charles Lyell; Hon. P. Pusey, M. P. from Berkshire in England; Hon. William Monsell, M. P. from Limerick County in Ireland; Right Hon. Alexander Macdonald of Dublin; Professor J. F. W. Johnston of Durham, England; Chevalier Bunsen, Prussian Minister in London; Professor Reissen of Bonn; Professor Dr. Sandberger of Weisbaden, &c.

Before I proceed to the details concerning particular schools, it may be well to state some general facts concerning them, such as the following:—

A few of them are founded and sustained by individuals, or private associations of gentlemen interested in Agriculture.

A far greater number are assisted, or sustained, by the government of the country.

Some consist of Professorships in Colleges or Universities.

Some depend upon Colleges and Universities for a part of their Professors.

The greater number are independent institutions.

Some are connected with primary schools.

We find among them three or four grades in the course of instruction.

In nearly all cases they have smaller or larger farms connected with them.

I shall now proceed to give an account of the Agricultural Schools which I have either visited, or concerning which I have obtained information. And here the difficulty will be, not to find facts, but out of the multitude that have fallen under my observation, to make an appropriate selection. The fact is, that the general peace that now prevails in Europe, has turned the attention of the people and of government to the arts, by which the ravages of a thirty years' war may be repaired; and among these arts, agriculture of course claims special attention. Says the French minister of agriculture and of commerce, in his admirable report on Agricultural Education,—"Since the mental activity of men no longer finds aliment on the field of battle, it is directed towards the labors and the conquests of peace. Everywhere the population has returned to industry, to the arts, to agriculture, to seek a remedy for the deep wounds made by war."

In giving details, I shall follow the order in which I visited the several countries.

ENGLAND.

The Agricultural Schools of England are of two grades, differing mainly in the amount of land cultivated, in the amount of apparatus and specimens, and in the extent of instruction. As yet the government have not, I believe, afforded them any pecuniary aid, and they are sustained by individuals, or by associated voluntary efforts. There is only one school of the higher grade, and that is called a college.

Agricultural College.

This is situated at Cirencester, about ninety-five miles northwest of London, on the oolitic formation, and adjoining the parks of Earl Bathurst, who is honorary president. The buildings are substantial, ample, and even elegant; the principal front being 190 feet. They include a dining hall, library, museum, lecture room, theatre, laboratories, class rooms, private studies, kitchen and servants' rooms, and offices with dormitories, one for each student. An elegant chapel has just been built. A forge, carpenters' and wheelwrights' shop are attached, as also a dairy and slaughter house.

Much attention is given to practical chemistry, the laboratory being on the plan of that at Giessen, and well adapted to manipulations. Nearly one half of the students become respectable practical chemists. The collections in natural history are not large. The students' private rooms seemed to me too small, and to look too much like the cells of criminals in a prison.

The farm attached, consists of 700 acres, some of which appeared to me to be rather poor land: a thin soil on limestone; but it is well cultivated. Upon it, besides the ordinary out houses, is a veterinary hospital.

There are accommodations here for two hundred students, but only fifty now belong to the school. Those residing in the building pay $355 annually; those who board elsewhere, $175. Formerly the school was open for the sons of the smaller farmers, but could not find support on that plan, and it was found, that if these attended, the wealthier classes would not send their sons. The price, accordingly, has been raised, and none but the sons of gentlemen, such as clergymen and wealthy laymen, now attend. None of the nobility send their children, although many give their money for its support.

John Wilson, F. R. S. E., to whom I am much indebted for giving me every facility for learning the character and condition of the school, is at present the principal and professor of the science and practice of Agriculture. With him are associated a professor of chemistry and chemical manipulation ; one of geology, natural history and botany ; one of mathematics and natural philosophy ; one of veterinary practice, and one of practical surveying and engineering. The instruction is mainly by lectures and subsequent examination, save that the pupils spend about half the day upon the farm, and most of them devote a good deal of time to practical chemistry. The instruction extends over three years, or six sessions ; the first year being devoted to *practical* agriculture. The courses of lectures, and their number, are as follows :

Science and practice of Agriculture, 3d, 4th, 5th, and 6th sessions, - - - - - - -	216	lectures.
Chemistry and chemical manipulation, 3d, 4th, 5th, and 6th sessions, - - - - -	152 to 172	"
Devoted to practical chemistry, - - - - - -	432	hours.
Botany, geology and zoology, all the sessions, - -	180	lectures.
Veterinary practice, " " - - -	180	"
Mathematics, 3d and 4th sessions, - - - - -	108	"
Natural philosophy, 3d, 4th, 5th and 6th sessions, -	72	"
Surveying and practical engineering, 5th and 6th sessions,	72	"

To give a better idea of the nature of these lectures, I quote below a few abstracts of one or two of the courses.

Science and practice of Agriculture. Fourth Session.

The Botany of Agriculture.

" The character and nature of the various agricultural plants, indigenous weeds, kitchen vegetables, fruit and forest trees, vegetable substances cultivated for, and used as food, and economic purposes."

The Geology of Agriculture.

" Condition and extent of surface soils : their influence on Agriculture,—local variations, conditions under which the different strata were deposited,—their composition,—mode of disintegration,—manner in which nature supported her own vegetation before man existed,—general surface geology of Great Britain."

The Physics of Agriculture.

" Meteorology, including climate,—mechanics, in reference to implements and machinery,—hydrodynamics,—the general laws of heat, light and electricity, as affecting the vegetable kingdom."

VETERINARY PRACTICE.

Division of the animal kingdom into four great groups.

1. Vertebrata.
2. Mollusca.
3. Articulata.
4. Radiata.

"Vertebrata,—1st class of,—mammalia,—considered and general structure described.

Character and habits of the 7th order,—pachydermata, including the horse, the ass, and the hog. Of the 8th order,—ruminantia,—including the cow, sheep, deer and goat—more particularly considered."

FOURTH SESSION.

"Structure of the animals in these two orders described, including, first, an account of the general anatomy and physiology of the elementary tissues of animals,—adipose,—vesicular,—epidermic,—cartilage,—bone,—teeth,—cellular, muscular, and nervous tissues,—fibrous, serous, and mucous membranes. Secondly, special anatomy of each animal,—the peculiarities of construction and the functions of its various organs and parts,—the head, neck, body, leg, foot,—points to be considered in the selection of animals, dependant on a proper knowledge of their anatomical structure."

FIFTH SESSION.

"Pathology and treatment of disease,—description of the various diseases incidental to the domesticated animals, as farcy, glanders, spavins, splints, curbs, capped hocks, sandcracks, cracked heels, blindness, &c. in the horse,—pleuro pneumonia, jaundice, red water, murrain, &c. in cattle,—rot, jaundice, dysentery, pneumonia, sturdy, &c. in sheep,—scrofula, measles, &c. in pigs,—their general causes,—mode of attack,—treatment, including medicines and operations,—neurotomy, bleeding, blistering, firing, &c.—precautions necessary."

SIXTH SESSION.

"The principles of breeding, rearing and feeding animals,—with examples, showing the benefit of their application. The treatment of stock,—stable management. The business and duties of the forge,—shoeing, clipping, singeing, &c. General remarks."

The collections in natural history are not large, but the library is well selected and rich in Agricultural works.

Students are admitted only upon the nomination of a proprietor or donor of $133, and of an age between sixteen and twenty. They "must at least be thoroughly well versed in the routine of a liberal education, in which they will be required to pass a strict examination before admission."

It will be seen, that the object of this college is not to fit men to become laborers on the farm, but to prepare them to be intelligent proprietors of farms, or to superintend those of others.

I was informed that this institution, which appears to be well organized, has met with a good deal of opposition from the lower and the higher classes, and Mr. Wilson feared that he might not be able to sustain it long in its present form. A rumor reached me before leaving England, that he had been obliged to include instruction in other branches of literature and science, in order to attract a sufficient number of pupils, but I cannot vouch for the truth of the report. From the unhappy operation, in this case, of *caste* in English society, and from the want of governmental patronage, we might reasonably anticipate, not merely embarrassment, but even failure, as facts respecting other schools will show.

Training Schools.

There are several schools by this name in England, in which agriculture forms a part, and an important part, of the instruction; but they embrace, also, the common branches of education. Usually a farm is attached to them. They form the second, or lower grade, of agricultural schools, and are mostly sustained by individual enterprise. I visited but one of these, of which a brief account only is necessary.

Agricultural Training School at Hoddesdon, Herts.

Mr. Haselwood is head master of this school, with whom are associated four other gentlemen as lecturers. They lecture upon the following subjects:

Chemistry and chemical manipulation.
Cattle physiology and pathology.
Botany, geology and natural history.
Practical mechanics and natural philosophy.
Practical surveying, levelling and draining.

Instruction is also given in the classics, and in the French and German languages. And separate classes are formed for those pupils who are not sufficiently advanced to attend the lectures and laboratory. The present number of students is thirty-five, but a small part of whom attend to agriculture. Opportunity is given for chemical manipulation, but not, as appeared to me, of a very elevated character. The specimens in natural history are few; but the philosophical apparatus and library, good. The farm attached, consists of three hundred acres, and includes a fine garden. The students are not compelled to labor abroad, but usually spend the latter half of the day in some out door pursuit, such as levelling, &c. There are two vacations of about six weeks each, one in summer and the other in winter; but there is not any fixed time for completing the course of study. The charge yearly is $140. I was pleased with the appearance of the premises, (it being vacation,) and am much indebted to the urbanity of the head master. The school was established in 1845.

Prayers are attended morning and evening, and attendance on worship on the Sabbath is required.

Agricultural Grammar School, Kimbolton.

I did not visit this institution, but received an account of it from the head master, Rev. John Thornton, M. A. It was originally a grammar school with a small endowment, and was reopened last year with a view to connect agriculture and its collateral sciences with literary instruction. Three other instructors are associated with Mr. Thornton, who give instruction in land surveying, the rudiments of agricultural chemistry, the veterinary art, geology, draining, entomology, vegetable physiology and mechanics. In a list of the half yearly examination, I find pupils in the following subjects: Arithmetic, mechanics, trigonometry, geography, Latin, Greek, French, history, algebra, Euclid's Elements, analytical trigonometry and agricultural chemistry. It was intended to have a farm attached, but the want of funds has prevented the execution of the plan. The present number of pupils is fifty, and there are three classes weekly manipulating in the laboratory. The cost to the pupil varies from $111 to $133. Hitherto it has been sustained in a good measure by subscriptions from public spirited gentlemen,—of the nobility, and others,—as is the Hoddesdon school; but now it rests solely upon the head master. It is "a grammar school," says Mr. Thornton, "where several subjects, not taught in schools generally, are taught, such as chemistry, botany, &c.

Several other institutions, similar to the above, exist in England. Near Kennington, not far from London, is the *Scientific and Agricultural Academy*, under the care of A. Nesbit and sons, which gives an education to those intended for the arts, manufactures and agriculture. There is a training school in the city of York, and I believe in other places, but I did not deem it necessary to visit many of them, as agriculture is only one of the studies pursued there.

IRELAND.

Agricultural education in Ireland presents itself under two phases. 1. In connection with the colleges. 2. In connection with the system of national schools.

1. *In connection with the Colleges.*

There are three colleges in Ireland called Queen's College ; one at Galway, one at Cork, in the southerly part of the Island, and the other at Belfast, in the northeasterly part. I visited only the latter, but became acquainted with the president of that at Cork, from whom I obtained an account of the system of instruction pursued there. Concerning that at Galway, I have no information ; but presume its organization to be similar to the others.

These institutions are, in fact, universities, having faculties in literature, science, medicine and law. At Belfast there are twenty professors, besides the president ; and at Cork, twenty-one. Among these chairs, one is a professorship of agriculture, filled by Edward Murphy, A. B. at Cork, and by John F. Hodges, M. D. at Belfast. The course of study and of lectures extends through two years, when the student receives, if accepted, a "Diploma of Agriculture." The courses of lectures embrace, in the first year, natural philosophy, chemistry, natural history, and the theory of agriculture ; in the second year, geology and mineralogy, history and diseases of farm animals, land surveying and the practice of agriculture. The details of these courses could be given, if wished for by the committee.

On the model and experimental farm, (comprising 180 acres at Cork,) and in the botanical gardens adjoining the colleges, and in connection with them, the students have an opportunity of becoming acquainted with the best kind of farm animals and machines, and with the manual and mechanical operations of practical agriculture, horticulture and arboriculture, being accompanied in their visits to see such objects and processes, by their instructors, as well as in various excursions of natural history.

Students who attend the agricultural lectures may be matriculated or non-matriculated. The former pay $33 each year, to the college ; the latter pay $9 for attendance upon any separate course of lectures. They also pay $3 annually for access to the library, which is well furnished with agricultural publications, to which the matriculated students have access without charge.

In each of these colleges are four scholarships of Agriculture, of the value of $67, two for each year. Candidates for these undergo certain examinations. For the first year, they must have passed the matriculation examination, viz.: in English grammar, composition and arithmetic, and then in the following subjects : English grammar and composition, the first four rules of arithmetic, vulgar and decimal fractions, involution and evolution, proportion and simple interest, mensuration, book-keeping, outlines of modern geography. For the second year, the examinations are in the general principles of heat, chemistry, mechanics and hydrostatics, elements of botany and zoology, theory and composition of manures, and feeding of farm animals.

Candidates for the diploma of agriculture pay to the college the first year, $33 ; for the second, $31. If they have scholarships, they pay only $20 the first year, and $18 the second.

The laboratory at Belfast college appeared to me to be fitted up in superior style, and to afford rare facilities for those students wishing to engage in manipulations.

For these facts I am indebted to Sir Robert Kane, president of Cork College, and to Professor Andrews, vice president of that at Belfast, with both of whom I had the honor and pleasure of an acquaintance, and to whose kind attentions I am much indebted.

These colleges have been opened only a little more than a year, and therefore it is not possible to determine how far agricultural lectures will attract students. At the first course in Belfast, I was told, only six or eight attended.

2. *Agricultural Institutions in connection with the National Schools.*

I would gladly give an account of the whole system of the National Schools in Ireland, for its features appear to me quite interesting, and to promise much for that country. I can only say, that in addition to the common schools, the system embraces model schools, training schools, industrial female schools, work-house schools, evening schools, and Agricultural Schools. These are all under the superintendence of a board of commissioners, appointed by government, who make an annual report. Dublin is the centre of operations, where model schools exist, with suitable and extensive buildings. There the superintendent of all the schools resides ; who, at present, is the Right Honorable Alexander Macdonald. Through the unsolicited kindness of Hon. Mr. Monsell of the British Parliament, I was introduced to Mr. Macdonald, and received from him all the aid I needed.

There are two grades of the Agricultural Schools,—the "Model Agricultural Schools," and the "Ordinary Agricultural Schools." The pupils in the latter, usually quite young, if distinguished, pass into the former, and are fitted to become teachers. In all these schools of every grade, literary instruction is combined with agricultural; and, indeed, the ordinary schools are only elementary schools, in which agriculture is taught.

Model School and Farm at Glasnevin.

The central and most important of the Model Agricultural Schools is situated at Glasnevin, in the suburbs of Dublin. Mr. Macdonald was kind enough to conduct me thither, and I had, consequently, a good opportunity for its examination. It is situated on a farm of 128 acres, of good soil, and the present head teacher is Mr. Donaghy, to whom I feel under great obligations for the time and attention he devoted to satisfy my inquiries.

This institution was established in 1838, and its grand object is to train up teachers for other schools; several hundreds of whom have already been sent out, and are spreading the knowledge here gained in other parts of Ireland. The present number of pupils is about fifty. But buildings are now in a course of erection for one hundred, at an additional cost of $24,930. The pupils receive literary as well as agricultural instruction. The principal lectures are on practical as well as theoretical agriculture. The mornings as well as evenings are devoted to study, but a large part of the day to labor. Twelve poor boys are placed here for instruction, for whose education and board the principal receives $1.75 per week. The boys also receive sixpence per week, (about twelve cents,) which the teacher told me was a sufficient stimulus to keep them at work. Most of the pupils, I should judge, are over twenty years of age. It was vacation when I was there, yet some thirty or forty had remained to work on the farm, and I very thankfully accepted an invitation to listen to an examination of the young men in the studies they had been taught. More than twenty cheerfully came in from the field, and without changing their dress, passed a very creditable examination upon the various principles of practical and theoretical agriculture, in connection with its associated sciences. I am sure that they cannot carry abroad such principles as they here presented, without doing immense benefit to impoverished Ireland.

On the farm, the principles taught in the school are practically illustrated. I walked over the fields, and have never, in any country, seen crops as fine, taken as a whole, of wheat, oats, beans, flax and potatoes. The oats would probably yield eighty bushels to the acre, and the potatoes bid fair to produce seven hundred bushels, the disease having not then shown itself. It is an object with the culturist to carry on the farm and keep it in a good state, without going beyond it for manure. A judicious rotation of crops greatly aids this object, and I believe that, ordinarily, manure is applied only once in four years.

The pupils have access to a good agricultural library, but I saw no collections in natural history, nor in any other department, indeed. The place, however, being only three miles from Dublin, the pupils can resort thither for instruction in natural history, and the inspection of specimens. There is a museum of economic geology there, which will ere long afford great facilities to pupils.

The pupils at Glasnevin are selected by the commissioners from the most deserving and talented young men in the various agricultural schools in the kingdom; so that here we might expect to find pupils of a high character. If they can succeed in extending the skill and productiveness exhibited on this model farm, throughout Ireland, I am confident we should hear no more of her population as starving.

The master of this school pays a rent for this farm, I believe as high as $22 per acre, increased to $35 by taxes and other expenses. Yet, according to Mr. Colman, he is able to make his labors profitable.

Larne National Agricultural School.

Of the smaller model schools, I visited that at Larne, about twenty miles north of Belfast, on the coast. This school has long been in successful operation. The Glasnevin school receives only in-door scholars, but that at Larne both in-door and out-door scholars; that is, such as board in the establishment and such as board out. Only a part of the pupils of the school, which is a national one, attend to agriculture. The present head master is Mr. M'Donnell, who seemed to me like his predecessor, Mr. Donaghy, now at Glasnevin, well qualified to unite the duties of a literary teacher with those of a practical agricultural instructor. The farm consists of only seven acres. Yet in 1848, he maintained on this small plot of ground, in the very best condition, three milch cows, two calves, four pigs, and one donkey, and raised besides 32½ cwt. of wheat, 28 cwt. of oats, and 24 cwt. of potatoes. The crops growing this year, which Mr. M'D. showed me, appeared unusually fine.

The in-door pupils pay $54 a year, including instruction and board, or if upon scholarships, only $22. The out-door pupils pay for instruction, $17 annually. The boarders work on the farm from 6 to 8, and from 10 to 12 A. M., and from 4 to 6 P. M. From 12 to 3 o'clock daily they study in the school room, in agriculture as a science as well as in literature; also, from 6 to 8 P. M., in an evening class under the superintendence of a teacher. They are not admitted under fifteen years of age, nor without a certificate of moral character. The course is of two or three years' duration, according to the age and acquirements of the pupils.

The agricultural instruction "embraces the principles of chemistry; the formation, nature, and difference of soils; the rotations of cropping best suited to such varieties; draining, trenching, and subsoiling, and the principles upon which their efficacy depends; house feeding of cattle, and its advantages; the constitution and properties of the different manures; the proper divisions of farms, &c., &c." To this is added a well grounded course of English education in reading, writing, arithmetic, English grammar, geography, book-keeping, mensuration, land surveying, gauging, geometry, trigonometry, algebra, and navigation.

Although it was out of school hours when I visited the establishment, Mr. M'D. offered to give me a sample of the proficiency of the pupils, and called in a few for this purpose. Although quite young,—not so old as fifteen,—they passed a very creditable examination upon agriculture, and some of the sciences, and the impression was strong upon my mind, that if large numbers of the youth of Ireland were thus instructed, from all classes in society, it must exert a great influence upon the agriculture of the country, especially when we know that they see the principles which they learn, fully and successfully carried out into practice upon the farms connected with the schools.

About one-quarter of the pupils at Glasnevin are Protestants, and the others Catholics, and this may be about the general proportion between the two denominations in the country. But such arrangements have been made, that each class receives religious instruction from clergymen selected by the parents or guardians. If the teacher of the school wishes to communicate religious instruction, he gives public notice of the time and place, and the pupils can attend or not, according to the wishes of their parents, or their own.

Other Model Schools.

Not less than twelve other Model Agricultural Schools were in operation in Ireland in July, 1850. Mr. Macdonald kindly furnished me with a list of them up to that date, as follows: Markethill, in Armagh county: Hollyrood, in Down: Carrick, in Fermanagh: Longhash in Tyrone; Sallybank and Belvoir, in Clare: Rahan, in King's county: Loughrea and Ballynakill, in Galway: Kile Park, in Tipperary: Bailieborough, in Cavan, and Dunmanway, in Cork. Besides these, building grants have been made by the commissioners in ten other places, viz.: Dunlewy, in Donegal: Bath, in Monaghan: Mount Trenchard and Tervoe, in Limeric: Ardfinnan and Derrycastle, in Tipperary: Woodstock, in Kilkenny: Leitrim, in Leitrim: and Glandore and Farraghy, in Cork.

The following are the essential conditions adopted by the commis-missioners, upon which they establish model schools :—

1. " The commissioners will take land, from eight to thirty acres, for the purpose of model farms, at a moderate rent, on a lease of at least 'three lives or thirty-one years.' "

2. The lease must contain a clause of surrender every fourth year.

3. The commissioners will not commence rent, nor enter upon the land, except the portion upon which the buildings are to be erected, until the 25th of March, or the 29th of September next ensuing after the completion of the works.

4. The commissioners will grant towards the building, a sum not exceeding **$1,776**, unless in cases where they may deem it desirable to provide two school-rooms. The remaining portion of the expenditure must be locally subscribed, and the amount of local contribution must be lodged in the Bank of Ireland, to the credit of the commissioners, before the works are commenced.

5. The buildings will be put up to tender; they are to be erected under the supervision of the architect to our Board, or of the clerks of works; and the grants will be paid by instalments on their reports.

6. The commissioners will furnish the dormitories and school-house.

7. The commissioners will supply (in the first instance) the necessary stock, farming implements, seed, &c., &c.

8. The commissioners will contribute **$33** towards the maintenance of two resident agricultural pupils, provided the pupils or their friends contribute a like sum.

9. Where one teacher only is required, the commissioners will grant **$44** a year for his services as agriculturist, in addition to his class-salary as literary teacher.

10. Where the farm consists of fifteen acres, or upwards, the commissioners will grant a salary to an agriculturist, not exceeding **$133** a year, and also to a literary teacher according to his class.

11. The commissioners will require the teacher or agriculturist to pay a moderate rent for the land, and all taxes, rates, &c., allowing him the profits arising from the farm. They will also require him to enter into arrangements for keeping up the supply of stock, implements, &c., &c., and for providing for permanent repairs.

12. The agriculturist will be required to conduct the operations of the farm according to the directions of the agricultural inspector, and must furnish accounts in the form prescribed by the commissioners. He must also submit annually to the board, to be laid before Parliament, a statement of the working and progress of the farm during the past year.

13. The commissioners, in consideration of the large amount of expenditure incurred by them, and the land being vested in them, deem it indispensable that they shall have the exclusive management of the Model Agricultural Schools; the right of appointing and removing the teachers and resident agricultural pupils; the latter to be selected from among the pupils of the National Schools in the district in which the model agricultural school is situated."

It is a most important fact to be noticed respecting all the Agricultural Schools of Ireland, and indeed of all Europe, that are sustained by the governments, (excepting perhaps in a few colleges,) that a farm, larger or smaller, is always connected with the school, so that the theories taught are there tested and exhibited in practice. Indeed, on the continent, in some places, unless the attached farm can be made to exhibit a state of cultivation fully equal to any around it, the government withdraws its support. In these facts we see that the objection so often urged in all countries against agricultural schools, that they teach mere theory, is done away with; for here, unless the instructors can show the truth of their theories in practice, they lose all patronage. Who can ask for a more severe test of theory than this?

Ordinary Agricultural Schools.

These are the elementary national schools, to which land, to the amount of two or three acres, is attached, and instruction in agriculture is given by the schools to those pupils who desire it. The only aid which they receive from the public funds is an addition to the master's salary of $22. "He pays the manager a moderate rent for the farm," say the commissioners in their report for 1849–50, "and receives the amount of the produce sold." These schools are, in general, working successfully, and have furnished satisfactory proof that literary and agricultural instruction can be practically united, without counteracting or encroaching upon each other. One inspector of Agricultural Schools, Dr. Kirkpatrick, has been active in the discharge of his important duties. All the existing Agricultural Schools have been visited by him once during the year, the majority of them twice, and several more frequently. In his report he observes,—"I feel gratified in expressing my strong conviction, that it is perfectly practicable, and eminently useful, to combine with the ordinary branches of a sound English education, as taught in our national schools, such an elementary course of agricultural instruction as shall prepare youths for the higher branches of agricultural science, should the opportunity of acquiring such knowledge be presented to them, and what is of still greater moment, shall teach them to avoid those grossly defective methods of farming, hitherto practised and still in too general use, throughout the greater part of Ireland." The commissioners add: "We are of opinion that the chief good that can be effected by us in the way of agricultural improvement, is by blending, in as many of our 4500 schools as possible, instruction in agriculture, and daily occupation in agriculture, with the literary instruction already given in those schools. Should the plan proposed by us be largely adopted throughout Ireland, improved agricultural knowledge and skill will be diffused throughout every part of the country and throughout the whole mass of the rural population. The boy taught in one of those schools will be enabled, in after life, to contribute his full share to the agricultural prosperity of the country, whether his vocation be that of a farm laborer, a small farmer, or a large farmer. He will, from his childhood, be taught to labor on the land, and to labor skilfully, to see displayed the rotation of crops, the application of manures, the management of cattle, the art of trenching and draining land. Every habit thus acquired by him, every kind of agricultural knowledge thus conveyed to him upon the limited farm of the teacher, will be equally serviceable to him, should he in after life become a large farmer, or if he never rise above the condition of a cottier."

According to the list furnished me by Mr. Macdonald, the number of these ordinary agricultural schools in Ireland, in operation last summer, was thirty-four, a list of which will be found in the synoptical table appended to this report. Many other applications for their establishment are under consideration, and they will no doubt be soon greatly multiplied.

This placing of agricultural instruction within the reach of the humblest classes of society, is a feature of the Irish system which gives it a great superiority to the English, especially as a model for our imitation. For were such schools among us confined to certain classes, and they the wealthy, it would produce little else than mischief. We can, however, more easily make this instruction accessible to all in our country, on account of the good education received here by the poorest classes. We can start from a higher level than can be done in any European country.

The whole amount paid by the Irish commissioners in 1848–9 (April to April,) for all the national schools, was $567,331. Of this sum, as nearly as I can ascertain, $11,000 was appropriated to the Agricultural Schools, a fact that shows the liberality of the government and their sense of the importance of this class of schools; and yet, we shall see that some of the continental governments are far more liberal. In Ireland the wages of instructors are quite low, as well as the board of the pupils; quite different from what is the case in England. In short, the system adopted there, appears to me to be in successful operation, and to afford many useful hints to those who are seeking to establish agricultural schools for the whole people.

The salaries paid to the regular teachers in the national schools, (including, I suppose, the agricultural,) varies from $67 to $133. Instructors of the model agricultural schools receive a gratuity of $44, and of ordinary agricultural schools, of $22.

In addition to the schools above described, I ought to mention that in the work-house schools, amounting to 111, where paupers are taught, instruction in agriculture is given.

Agricultural Schools sustained by private Associations.

Templemoyle School.

Besides the schools under governmental control, and fostered by the public funds, a few others exist in Ireland. The most distinguished of them is at Templemoyle, near Londonderry, in the north part of the island. It "was founded in 1827, by the North West of Ireland Society, on the plan of Fellenberg's School at Hofwyl, in Switzerland. At first, in imitation of Fellenberg's plan, two branches of the school were established; one for the sons of the wealthy, the other for the poor. But a short and expensive trial convinced the society that they could not sustain both establishments; and they wisely gave up the more expensive one, and confined their whole means and attention to the more practical and industrial school for the poor. This is probably one grand reason why the school at Templemoyle still lives, and has done a great deal of good, while that at Hofwyl has been abandoned.

The pupils in the Templemoyle school are children neither of the wealthy, nor of the very poor, but rather of "honest and industrious farmers;" and from the education which they receive, they are prepared "to act, either as the managers, or stewards, of the landed proprietors; as improving farmers on their own account; or as masters in other schools of a still more industrial character." Hence a good deal of time is devoted to scholastic training, as well as instruction strictly agricultural. Botany and chemistry are introduced, as well as lectures on the veterinary art; and the culture of forest trees, a nursery, and a flower and vegetable garden.

The internal regulations of this school as to hours of rising and retiring, labor and study, week days and Sundays, diet, &c., &c., appear to be judicious, and adapted to the strictest economy. But as they are given fully in Mr. Colman's reports, I judge it unnecessary to copy them.

The farm contains 169 acres, a part of which is cultivated upon the four shift, and a part upon the five shift, rotation system, as it is called. The annual charge to a scholar is reckoned at about $44. Up to 1838, the contributions and payments from pupils and others had amounted to $35,520. This had all been expended, and a debt of some $2,000 was due from the school, but was gradually diminishing, and the circumstances of the school have been growing more favorable, and its influence increasing.

Those who had left the seminary previous to September, 1838, were distributed as follows:

29 are employed as land stewards,
2 as assistant agents,
5 as schoolmasters,
1 as principal of an agricultural day school,
8 as writing clerks,
6 as shop-keepers,
1 as civil engineer,
2 as assistants to county surveyors,
124 are employed at home in agricultural pursuits,
32 have emigrated to America, the West Indies and Australia,
39 left the seminary since Sept. 1835, not having remained twelve months.

Most of the preceding facts have been obtained from Col. J. E. Portlock's Report on the Geology of the County of Londonderry, &c., Dublin, 1843.

At Brookfield, twelve miles from Belfast, a similar school was a few years since sustained by the Society of Friends, with a farm of fifty acres. Here the common branches of learning, as at Templemoyle, are taught, along with agriculture: such as "reading, writing, arithmetic, geography, the catechism and scripture history." There is only one male teacher with a female assistant, who are both under the direction of a man and his wife, who act as master and matron. Forgetfulness prevented me from making enquiries respecting this school when at Belfast, so that I can say nothing as to its present condition.

SCOTLAND.

Although Scotland has, within the last half century, been distinguished for her progress in practical farming, she has no agricultural schools, save a professorship of that department in two of her universities, viz.: one in Mavischal college in Aberdeen, and another in the University of Edinburgh. I have understood that the former has not succeeded in attracting many students, although filled by an able chemist, Dr. Smith, and I am unable to state its present condition. But having looked into this department of the university of Edinburgh, and having received some statements from the gentleman who fills the chair, Professor David Low, I will give a brief account of it.

Several rooms in the university buildings are devoted to this department. There we find exhibited, numerous models of agricultural implements, and of buildings, chiefly barns and outhouses, for the farm. Also seeds and dried plants, marls and soils, chemical substances, a few skeletons and anatomical drawings, and, more particularly, as many as a hundred fine drawings, mostly as large as life, of the best breeds of domestic animals. These drawings are chiefly given by Professor Low, in his two works on the domesticated animals.

Professor Low has been so obliging as to give me on paper a number of facts respecting his professorship, and he has added some remarks, which may be of use to Massachusetts should she establish agricultural schools, and therefore I give the whole in his own words.

EDINBURGH, 19*th August*, 1850.

DEAR SIR,—I have not published any syllabus of my course of lectures, but the arrangement of the more practical branches is the same as in the last edition of my Elements of Practical Agriculture, which work, indeed, was mainly designed as a text book for the course. Another branch is added, of importance in this country, namely, the management of landed property, comprehending the subject of the buildings and other appendages of the farm, of leases, and generally the relations between landlord and tenant.

The students consist chiefly of the sons of tenant farmers,—proprietors of land, or their sons,—agents having the charge of landed property,—and a mixed class frequently from other countries, who are desirous to obtain a knowledge of the system of agriculture, as it is pursued in this country.

The younger students generally fill up the period of the session by attending other classes in the university, or of private teachers, chiefly chemistry and natural history, and several of them likewise attend the course of lectures delivered in this city by Professor Dick, on veterinary surgery.

I have found by experience, that it is not necessary to have a farm in connection with the Chair. All the essential points of practice being previously explained, the students are prepared to enter upon the study of the subject in the fields. To this end, they usually board in the house of some respectable farmer for such a period as suits their convenience, not less, I recommend, than twelve months, so that they may see the operations of an entire season. There are numerous respectable farmers, both in the south of Scotland and in the north of England, who are in the habit of receiving pupils and instructing them in the different branches of their profession. The usual charge is £100 a year ($444), and an allowance, where the pupil keeps a horse, of £25 more ($111). In the United States a less artificial system of management of land suffices than in this country, and I doubt if the sending young men to study on the farms of the country itself would have the same beneficial effects as in this.

The means of communication are now so easy, that I think it would be worth the while of any American gentleman especially desirous of studying agriculture in a good school, to settle himself in this country for twelve months or more, under the tuition of a respectable and skilful practical agriculturist.

I believe it is quite indispensable that a teacher, whether by means of academical lectures or otherwise, be himself thoroughly acquainted with the practice of the farm. He will otherwise be unable to give that kind of instruction to the students which will really be of use to them on their farms. If, therefore, you establish agricultural professors in any of the American universities, I would recommend that the intended professor be sent to this country, so that he may be rendered thoroughly acquainted with the best practices of agriculture; and the person chosen for this office should be one previously sufficiently instructed in the different branches of physical science which have relation to agriculture. I think if this were done, you would have no difficulty in getting professors for your universities who would attract students to their lectures. You ask me if members of the university attend the class. A few of my colleagues have done so, but generally, persons only do so who have some connection as proprietors, tenant farmers, or otherwise, with landed property.

I enclose a list of the paintings of animals in the museum, the subject of which is treated of in detail in my work on the domesticated animals.

The management of landed property is of less interest in your country, but I beg you to accept of a copy of my work on this subject, which I think will enable me to exhaust the subject of your inquiry.

I am, dear Sir,

Yours very faithfully,

DAVID LOW.

A most important fruit of agricultural schools and professorships in Scotland and Ireland has been the preparation of books on the subject. In Ireland several have been published which are so well adapted to aid in elementary instruction, that they have been a good deal used in England. Professor Low has published several large and important works on this subject, to some of which, allusion has been made above. Such works cannot but produce important effects upon the progress of agriculture; and the multiplication of schools and professorships of this sort, cannot fail to increase the number of works calculated to diffuse widely the true principles of husbandry.

Both in England and Scotland, and indeed in Ireland also, agricultural societies have been an important means of advancing practical farming. Some intelligent gentlemen, whom I met abroad, were even of opinion that these societies were all that is needed, and that schools would be superfluous. But men do not so judge in respect to other departments of knowledge. Societies have long been in active and successful operation in the various physical sciences; but they are not thought to render schools and colleges unnecessary. The societies aid the schools, but the schools must train up men to be efficient members of the societies.

And here let me remark, that in my opinion, schools in addition to societies are most needed in countries whose agriculture is not in the best state. England and Scotland can probably better dispense with schools of this sort than such a country as ours, where so much remains to be done to bring our farming up to the standard of those countries,—or rather to a higher standard.

But my special object in alluding to societies in this connection, was to call attention to the fine collections belonging to the Highland Agricultural Society, and deposited in Edinburgh. Here we find two beautiful rooms and well lighted, one above the other, well filled with the following collections:

1. Drawings of domestic animals.
2. Agricultural implements, (small models.)
3. A great variety of seeds.
4. Grasses and grains of full length, fastened to the wall.
5. Blocks of wood, used in the arts, planed and varnished.

6. Specimens of rocks, soils and marls.
7. Specimens of insects injurious to vegetation.
8. A small green house.

These rooms certainly afford a fine model for an agricultural museum, and cannot but be very instructive to an intelligent farmer. They are just such collections as are appropriate for an Agricultural School.

CONTINENTAL EUROPE.

It is now more than half a century since Agricultural Schools were first established on the continent of Europe. The two earliest, started in 1799, were those at Hofwyl, in Switzerland, and at Krumau, in Bohemia. The former no longer exists, but the latter is still in operation. From that time to the present, individuals and societies have put such institutions into operation in various countries. Within a few years, however, there has been an extraordinary movement among the governments on this subject, and such schools have been rapidly multiplied. I have visited a few of them, but derive a large part of my facts concerning them from an able report on this subject, by the French minister of agriculture and commerce, published during the present year. The great resemblance between these schools, in their general character, will enable me to be comparatively brief in my descriptions.

FRANCE.

"The institutions of society, like the laws," says the French minister, "which satisfy the real wants of the public, do not spring up spontaneously. Their history shows us that they are always called for beforehand by the public will, preceded by isolated efforts and partial attempts, up to the auspicious moment when the legislator, constrained by the general sentiment, and enlightened by the experiments already made, finds, so to speak, the basis and the materials all ready for the new edifice, called for by the new wants of the country."

The earliest effort in France, and indeed in Europe, so far as I can learn, to establish an agricultural school, was made by M. L'Abbe Rosier, in 1775. He called it "a Plan for a National School of Agriculture in the Park of Chambord;" a place built as a retreat for the court in time of danger, by Francis I, at great expense. It was afterwards presented to Marshal Saxe. It had fallen into neglect, and Rosier's plan was to establish a school there for teachers, as well as practical agriculturists; not very different, indeed, from the plans now extensively adopted in France. Rosier first proposed it to the French minister, Turgot, in 1775, who received the plan with favor, but was driven from his place before he could carry it into execution. In 1789 it was offered to the National Assembly, but received coldly, and when, in 1791, Rosier tried to recover his manuscript from the national archives, it was found that it had disappeared. It was at last found to have been carried into Spain, from whence it was recovered. Meanwhile Rosier was killed in the seige of Lyons. Yet his patron, Le Comte de Neufchateau, developed his plan, and subsequently tried to interest Bonaparte in it. But he, finding that Chambord was originally a military post, chose to reserve it for that purpose, so that France lost the honor of being the first European nation to establish an agricultural school.

It is not improbable, however, that this plan of Rosier might first have suggested to Fellenberg of Switzerland, the idea of the Hofwyl school, which was established in 1799, but has lately been discontinued.

There was one feature of this plan of Rosier's, which should perhaps be mentioned, since it might be useful at this day, and indeed it has been adopted in some of the English schools, at least. To each pupil was assigned a small plot of ground, with all the instruments necessary to its cultivation, and he was to cultivate it according to the methods adopted in the region from whence he came; receiving himself the income.*

We now proceed to mention some of the earliest instances in France in which agricultural schools were actually established, although some of them were at length discontinued for want of funds.

As early as 1818, M. de Domsbasle rented an estate at Roville, of 375 acres, at $2,000, to make it a model farm, and a place of education. He was obliged to resort to a subscription of 45,000 francs, to put the establishment into operation, which was done in 1822. And for six years it flourished, and did much to awaken in France a taste for agricultural studies, and a reform in the defective modes of culture that prevailed. The results were published in a periodical, entitled *the Annals of Roville*, which diffused, far and wide, more enlightened views of agriculture than had prevailed. M. de Domsbasle, however, found that his funds were insufficient to sustain the establishment in a

* I cannot understand why the French minister, in his able report on Agricultural Schools, wherein he describes many other pioneers in this work in France, (as we shall shortly see,) should pass Rosier in silence; a man who presented a sagacious, and for the most part, a judicious plan, for such schools, almost a half century earlier than any whom he does describe. Neither do I understand it, that while he speaks in the highest terms of the agriculture of Great Britain, he should mention as existing there, only one agricultural school,—that at Cirencester,—whereas, at the time he wrote, not less than seventy such schools existed there, as the tabular view, annexed to this report, will show. The facts respecting the Abbe Rosier, I have derived from Portlock's Report on the Geology of Londonderry, &c., p. 711.

healthy state. The revolutions in France, also, brought on new embarrassments, and he resorted to the government, which at last gave partial aid, but not enough to give permanency to the school; and after a twenty years' struggle, finding his own constitution, and that of his wife broken down, he abandoned the enterprise in 1842. The example, however, doubtless did good, and led to the establishment of other schools, on a securer foundation.

The Royal Agronomic Institution of Grignon was started in 1827, through the impulse given by the Roville school. This place is about twenty-five miles southwest of Paris, and twelve miles from Versailles. Favored above Roville in soil, capital, and patronage, it has had a more fortunate progress, and never was in a more prosperous condition than when I lately visited it. With a farm of 1250 acres, and receiving yearly from the government about $12,000, it seems to be doing much for the agriculture of France. But I will delay a little a full description of this school upon its present organization, in order to speak of some others, started by individual enterprise, though aided by the government.

The Institute of Coetbo was one of these. Its objects were to enable agriculture in France to attain the degree of perfection it had reached in Great Britain, in Flanders, and in Switzerland; to collect together from all the departments of France, pupils, who might return, amply prepared to teach a better system of agriculture; to raise up a corps of *agricultural engineers*, prepared for any important enterprise in this department; and finally, to furnish a wide field for experiments on the amelioration of farms, such as individuals could not undertake.

The course of instruction for these objects was full, perhaps too full for a beginning, and without very large governmental patronage. For ere long the institution became so much embarrassed in its pecuniary concerns, that it was given up; not, however, till it had produced a good impression on the public mind, and made it more and more manifest that such institutions were needed by the country, and that it only required a more liberal patronage from the state to give it permanence and extensive utility.

When the Roville school was given up, one of the pupils went into Brittany, and took an estate of 1250 acres of very barren land, two-thirds of it being covered with heath, and by applying the scientific principles learnt at Roville, he reclaimed the soil to such an extent as to excite public attention, and lead at length, in 1833, to the establishment of the school of Grand Jouan, which still exists, though like that at Grignon, its organization is somewhat changed, as will be explained in the sequel. The success of this school, in such an unpromising spot, is one of the most striking examples of the importance of such institutions.

In the southern part of France, another interesting experiment has been performed by another individual, M. Neviere, who caught his spirit from the Roville school. Between the Rhone and the Saône, not far from Lyons, there exists a region, of sixty-seven square leagues, which scarcely sufficed to sustain a scattered and miserable population, the mean length of whose lives did not exceed twenty-five years. The soil and climate are, indeed, naturally good ; but it is periodically inundated by 1600 or 1800 marshes and ponds, which bring pestilence with the overflow. M. Neviere purchased 1600 acres, and in two years succeeded in draining all the ponds, and introducing successful cultivation, with the most marked benefit as to health. Before the drainage was finished, the per centage of his domestics, sick with fevers, was as high as twenty, from June 15 to October 15. After the drainage, the proportion fell successively to six, five, three, and finally to one-half per cent.

The school established in this region by M. Neviere, was at Saulsie, where was a model farm. Its objects were special ; adapted to the region where it was placed. These objects were to show how operations could be conducted on a large scale, such as the raising of forage for cattle, and of the cereal grains. He probably laid his plans too large, at least for the limited aid he received from the government, and for a time the institution became almost extinct. But in 1848 the government remodelled it, and it has since been in successful operation, on a plan to be shortly explained.

The four institutions which I have now noticed, were intended to teach the higher principles of agriculture to proprietors of farms, to large farmers, and to managers of farms ; in a word, to prepare the officers of the agricultural army. But the sub-officers, and rank and file, were not forgotten ; that is, the small cultivators, the master servants, and servants of the farm. To some of the larger institutions were attached schools for the sons of the peasants, and for orphans, as it is now at Grand Jouan, but in no other school. Separate establishments were also got up, under various forms ; such as farm schools, rural asylums, agricultural colonies, penitentiaries, model farms, farm schools, and schools of agriculture. Their object was to teach the elements of agriculture and show the pupils how to labor in the most judicious manner. The government favored these projects of individuals and charitable associations. But the unavoidable difficulties attending every new enterprise pressed so heavily upon them, that all of them might have failed had not the government come more efficiently to their aid. The expenses of instruction were supplied from the national treasury, and the owner of the farm did the best he could with the farm. Not less than nine of these schools were thus continued in operation till 1847, when many new ones were begun, so that they amounted to twenty-five at the close of that year. About that time the government took new and important measures in relation to agricultural education, whereby the number of the schools of all grades has been very much increased, and the whole system placed on a more substantial footing.

This brief history seems to me important, chiefly because it records the failure of a number of agricultural schools, a point to be carefully looked at by those who are deliberating whether to attempt establishing similar institutions in other countries. The French minister makes some remarks on this subject worthy of a translation.

" When we turn our attention," says he, " to this first phase of agricultural instruction in France, it is trying to see the greater part of those establishments, which endeavored to rise, yield to the difficulties that met them at the outset. But reflection shows us that in this respect agriculture meets the fate of all human institutions, which arise in new conditions, and must pass through the changes of progress. The first who engage in them exhaust themselves in opening a path, so that those who follow may find it cleared, and may there gather the fruit of the sweat and labor of those who have gone before them. Rarely does he who opens the path arrive at the end of it, so that the history of discoveries, of ameliorations, and of advancement, is only the recital of the trials and the ruin of those to whom humanity is most indebted."

" But whatever may have been the want of success of the first agricultural establishments, they have notwithstanding contributed no less to the improvement of agriculture, and to the movement, which, at the present day, has advanced it in its progress. Their reverses ought not to make us forget their services ; for these reverses are the source from whence profitable lessons may be derived. From having seen their struggles with difficulties, the administration (of government) have been led to see in what their organization was defective, and to take advantage of their trials."

" It is one of these advantages to the farm schools and the institutes, that they have thus learnt the dangers and abuses to which such institutions are liable."

" From hence, solicitude has been awakened as to the danger of destroying, in the pupils of the farm schools, the austere and simple habits of life formed at home, and as to the precautions necessary to meet this danger, and to restrain within wise limits their course of instruction."

"From hence, too, we have learnt the difficulty of collecting in some localities a sufficient number of pupils for farm schools."

"From hence, also, light has been thrown upon the institutes, (higher schools of agriculture.) We have seen from the first that the farms of those first establishments did not belong to the state ; that we were obliged to take for a director the owner of the farm, not always the best man. Hence, in some cases, the teaching was not sufficiently experimental, because the interest of the proprietor did not allow of the expense necessary for successful studies and experiments in agriculture. Still further, the state, after having been at much expense, were liable to be disappointed by the sickness or reverses of fortune of the proprietor."

"Enlightened by all this experience, the administration were able to offer such a project for a law concerning agricultural instruction, as would obviate the embarrassments and dangers which had injured or ruined some schools, &c."

The law above referred to, which forms the basis of the present organization for agricultural instruction in France, after having been proposed by the minister, was thoroughly discussed by the National Assembly, and finally adopted on the 3d of October, 1848. A translation of it will be the best means of understanding the character of the agricultural schools in France, for all the old schools were remodeled to conform to this new decree.

Law voted by the National Assembly of France, Oct. 3d, 1848: The vote stood 579 in favor, and 100 against the law :—

Preliminary Arrangements.

Article 1. Professional instruction in agriculture constitutes three degrees.

In the first degree are embraced the *Farm Schools*, for giving elementary practical instruction.

The second degree embraces the *Regional Schools*,* (*les écoles régionales*) where the instruction is both theoretical and practical.

The third degree embraces a *National Agronomic Institute*, which is a higher normal school of agriculture.

* I have thought it better to anglicise the French word here, than to translate it. *District* school would be quite proper, but in Massachusetts it designates the primary schools in a town, and would be so understood if applied here.

Article 2. The professional instruction in agriculture is at the expense of the state in its different degrees.

SECTION 1. OF THE FARM SCHOOLS.

Article 3. The farm school is a rural enterprise, conducted with ability and profit, in which apprentices are chosen among the laborers, and admitted without charge, execute all the labor, receiving at the same time, as a remuneration for their labor, an agricultural instruction essentially practical.

Article 4. In each of the departments of the republic, there shall be established at first a farm school. This organization shall be successively extended to each *arrondisement*, (subdivision of a department.)

Article 5. The salaries and wages of the instructors (*personnel enseignant*,) are paid by the state. The state also determines the price of the board, which, joined to the labor of the pupils, is allowed to the director to indemnify him for the cost of food and other expenses, occasioned by the admission of the apprentices.

Article 6. Each year the treasury distributes prizes to the farm schools. They are divided every year to all the children according to their merit, under the name of *pecule*, (stock of money,) but they are not paid over till the termination of the apprenticeship.

SECTION 2. OF THE REGIONAL SCHOOLS, (*les écoles régionales*).

Article 7. France shall be divided into agricultural districts, (*regions culturales.*)

In each district there shall be a regional school.

The *Regional School of Agriculture* is an enterprise for making experiments, and, at the same time, a model for the region to which it appertains.

Article 8. The pupils received into the regional schools are either those having scholarships, or those who pay for their board.

Article 9. The scholarships established in the regional schools are given, after consultation, one-half to the pupils of the farm schools of each agricultural district, and the other half to those persons who present themselves as competitors.

Article 10. The better pupils of the regional schools, who do not enter immediately into the national agronomic institute, may be placed at the expense of the state as licentiates, near the farm schools, and other agricultural establishments, public and private.

The duration of this residence is two years.

The licentiate assists the director in his labors, becomes acquainted with the practical part of the administration, and completes his agricultural education as *chef d'exploitation*, (chief of the farm enterprise.)

Article 11. The regional schools are also experimental farms.

The experiments and their results shall receive the greatest publicity.

SECTION 3. OF THE NATIONAL AGRONOMIC INSTITUTE.

Article 12. The National Agronomic Institute shall be established on the national estate at Versailles.

Article 13. The course of instruction of the national agronomic institute, shall be gratuitous and public.

Nevertheless, the state sustains there forty scholars.

Each year ten scholarships are given, upon consultation, to graduates of the regional schools; ten others are reserved for all the competitors who present themselves.

Article 14. Each year the three first scholars of the institute receive a *supplementary mission of studies.*

This *mission* extends to three years; it is open for Frenchmen and for foreigners.

Article 15. The national agronomic school takes also the experimental character bestowed upon the regional schools.

The experiments shall be made public, as directed in article 11.

SECTION 4. GENERAL ARRANGEMENTS.

Article 16. The functions of a professor in the regional schools, and the national agronomic institute, shall be given in council.

Article 17. The schools and the national institute shall be managed solely on account of the state.

Article 18. (Relates to local matters and need not be given.)

Article 19. Each year an account shall be rendered to the national assembly of the manner in which this law is executed.

Article 20. This law shall be executed by the statutes of the public administration and by the decrees of the minister of agriculture.

Six months after the adoption of this law, the national assembly, after a good deal of discussion, voted the sum of 2,500,000 francs ($460,000,) to carry its articles into execution for the year 1849. This was the sum recommended by the minister in his budget.

It remains that I should give a more detailed account of the three classes of agricultural schools which have thus been established in France, and to describe at least one example of each class.

Farm Schools.

I have already defined these schools to be a rural enterprise, conducted with ability and profit, and in which the pupils perform all the labor, receiving at the same time, as a remuneration, instruction in agriculture, essentially practical.

They are established for two objects: first, to furnish good examples of tillage to the farmers of the district. Secondly, to form agriculturists capable of cultivating intelligently, either upon their own property, or that of others, as farmers, leaseholders, or managers, or to become good rural assistants, servants upon the farm, leaders of manual labor, or overseers of cattle and horses.

It is the intention of government to establish at least one of these schools in each of the eighty-six departments of France; but as yet only seventy have been put into operation. They are placed as near as may be to the centre of the department, and in a spot where the soil, situation, &c., are similar to the general condition of the region. They have annexed to them, nurseries, and collections of fruit trees, and gardens. The buildings for the accommodation of the school are constructed in a plain and substantial manner, and so as to conform as much as possible to the general character of buildings in the district. The director is chosen of preference from among the farmers or proprietors of the district, whose farms are conducted in the best manner.

The instruction is, as far as may be, practical, and given even in the field, where the pupils labor, in the stables, and the sheep-folds.

The officers and head men of the establishment, with their salaries, are the following:

A *director*, having the general oversight. Salary 2,400 francs, ($442.)

A *chef de pratique*, chief of the practical operations of all sorts on the farm, 1000 francs, ($184.)

An *overseer of accounts*, who teaches the mode of keeping farm accounts, considered in Europe as a most important part of agricultural education. Salary, 1000 francs, ($184.)

A *nursery gardener:* salary, 1000 francs.

A *veterinary surgeon:* salary, 500f, ($92.)

Some other leaders, according to the locality, &c., such as shepherds, cheesemongers, irrigators, silk growers, &c.

The school is open to young men, particularly those from country families, whose age is at least sixteen, who have received an education in the primary school, and who have a good constitution. The number admitted varies with the extent of the country. If less than twenty-four apply, it is thought not best to continue the school. The actual number attending the schools varies from twenty-four to thirty-two.

The pupils all labor as workmen receiving wages. Three in each school are confined to the gardens and nurseries, so as to become gardeners. The number should be great enough to carry on all the operations of the farm, which is considered an essential feature of this sort of school. Board and instruction they receive gratuitously in return.

The course extends through three years. The first year, they attend to simple manual labor; the second year, they have charge of the animals; the third year, they have the oversight of various operations.

Provision is made for religious instruction, adapted, as in Ireland, to the peculiar religious condition of the country.

The hours appropriated to study are devoted, 1. To the arrangement of the notes the pupils have taken during the instruction from the different leaders; 2. to reading a manual, or book, of elementary agriculture; 3. to lessons given by the overseer of accounts, on the elements of arithmetic, book-keeping, surveying, &c. The time devoted to study is less in summer than in winter.

Besides paying the board of the pupils, the government allows to each one, as an encouragement, the sum of $14 yearly. One part of this is intended to cover the expense of the maintenance of the wardrobe (?) [du trosseau.] The other part forms a fund, to be distributed among the pupils each year, in proportion to their zeal and good conduct. These prizes are awarded, but not paid over, till the end of three years. He who quits the school, or is dismissed before that time, loses these prizes.

Another prize of 400 francs is awarded to the pupil most deserving, at the end of the three years' course.

The director works the farm at his own risk. He is obliged so to conduct it as to afford the best means of instruction to the pupils, to submit his books and accounts at any time to the examination of government, to send annually to the minister a full account of the state of the school, and to publish a full account each year of his operations, his success, or his failure. Moreover, if the farm be not conducted so as to afford a net produce comparatively equal to that furnished by other farms in the same region, the patronage of the government is withdrawn.

Instead of going into the details, as I might do, of the state of each of these schools, I have thrown together, in the table attached to this report, all the facts respecting them within my knowledge, which I judge important to the purpose of the committee.

Regional Schools.

I have already given a history of several of these schools of the second grade, but in their present organization they deserve a fuller description.

They have three objects in view :

1. To form enlightened agriculturists, by teaching them the principles of agriculture.

2. To offer an example, or model, of practical agriculture of a high order, and advancing.

3. To make experiments for improving the cultivation of the soil.

The instruction in these schools is of a much higher order than in the farm schools, and is adapted not to prepare laborers on the farm, so much as men to direct agricultural affairs. The farm connected with the school is expected to present an enlightened system of culture, and to adapt that culture to the wants and peculiarities of the district in which it is situated. The director, also, is no longer a farmer, or proprietor, laboring at his own risk, but an agent employed by the government, and accountable to them, and subject to their direction.

The instruction is both theoretical and practical, embracing the following six professorships :

One professor of rural economy and legislation.

One of agriculture.

One of zootechny, or the economy of animals.

One of sylviculture, (cultivation of forest trees,) and of botany.

One of chemistry, physics and geology, applied to agriculture.

One of rural engineering, (irrigations, rural constructions, surveying, &c.)

It may be well to add a brief explanation of the nature of the above professorships.

The course on rural economy and legislation describes the relation between rural productions and the public revenue, as well as the different branches of industry. It shows what circumstances are favorable or unfavorable to such or such a system of cultivation, or to such or such a speculation in animals, or vegetables, according to the situation of the lands, the facility of communication, and demand for the products by the people of the surrounding country. The course embraces also rural legislation.

The course on agriculture embraces the study of the soil, of manures, of instruments of tillage, of different cultivated plants, an estimate of the different modes of culture, and the theory of the distribu tion (rotation ?) of crops.

Zootechny treats of the production and amelioration of animals. The professor gives at first some ideas of anatomy and physiology generally, and then treats, in a practical way, of the raising of domestic animals, of their support, of their amelioration, of their hygiene, and their production.

The professor of sylviculture and botany gives first, a summary sketch of vegetable physiology and botany applied to agriculture. He teaches the subject of sylviculture (cultivation of woods,) and of forest economy, with special reference to the training, working, and preservation of the forests of individuals and the communes.

The professor of chemistry, physics, geology, &c., has a wide field, as his titles show. His chief object is to take those views of the sciences named which bear directly upon agriculture.

The professorship of rural engineering embraces geometry, mechanics, and linear drawing, as applied to rural architecture, to the construction of agricultural instruments, and particularly to irrigations.

To second the lessons of the professors, an equal number of tutors are appointed. Their duties are to explain in private, to the pupils, whatever is obscure or difficult in the oral instruction. They also see that notes are taken of the lectures, &c.

Each school has its library, its philosophical and chemical cabinet, adapted especially to agriculture, its agronomic museum of geology, zoology, botany, and agricultural technology.

The pupils have an opportunity of witnessing on the farms connected with these schools, all the important agricultural operations, also specimens of the best breeds of animals, and the mode of taking care of them, and using them ; and they engage personally in all the important operations connected with husbandry, so as to know how to conduct them in after life.

The number of scholars admitted is fixed by the government, and varies at the different schools. The price of board is 750 francs, ($138.)

The state furnishes several scholarships to each school. Half of them is given to the most deserving of the pupils from the farm schools, placed at the regional schools. The other half is divided among the scholars who are the most distinguished, after six months' trial, for their labor and conduct. Scholarships from the national agronomic institute, are also given to those most successful in study and conduct.

Towards the close of the third year, examinations are held, and to those who sustain them, diplomas are given, and the way is laid open for their admittance to the national institute.

To these schools a farm is always attached, for the purposes already indicated. Also, a manufactory of agricultural instruments, a *magnanery*, (establishment for silk,) a *feculary*,* (place for preparing liquid manures,) distillery, oil mill, dairy, sawmill, &c.

The head men on the farm are essentially the same as those already described as connected with the farm schools.

Regional School at Grignon.

As I visited this school, a few words more concerning it may be proper. It lies about twenty-five miles southwest of Paris, and twelve from Versailles, and may be reached by taking a railway to the latter place, and a carriage from thence. The country passed over is very fine, in an agricultural point of view, although I should not judge the farm surrounding the school to be unusually well located. I was accompanied in my visit by Mr. Morey, of Massachusetts, who kindly acted as my interpreter. We found M. Bella, the polite and intelligent director, ready to answer all our inquiries, and to show us the establishment.

I have already stated that this school, commenced in 1827, had lately been reorganized, according to the law of the national assembly of October, 1848; yet its plan of instruction and general character remain essentially the same. It is intended to educate young men, of good families, to become directors and stewards of estates, or to manage judiciously their own estates, and not to learn men to become themselves laborers. The pupils are required, however, to take part in all the operations of the farm, but not as a daily and irksome task. They take turns in attending to the different operations; one week in laboring abroad, the next in taking care of the cattle, &c., and during the third year they are made sub-overseers of different processes. The number of professors at present is six, and of course an equal number of tutors, and other head men, for practical operations. About seven of the instructors reside on the ground; the rest are gentlemen from abroad, some I believe, from Paris, who give courses of lectures. All the expenses of instruction are paid by the government, who, in fact, have a full control of the school, and of the farm, consisting of about

* These are merely anglicised words from *Magnainerie* and *Feculerie*.

1,200 acres. I noticed a few tolerably good collections in natural history,—a manikin, and some skeletons of animals,—but the collections are not large. That in geology is the best; the surrounding country, being situated in what is called the Paris basin, furnishes many interesting specimens. But some of the professors bring specimens along with them to illustrate their lectures.

As we passed over the farm, and through the buildings, in the house, and the barns, I observed that the pupils,—most of them at least twenty years of age, and some much more,—wore a neat frock, as did also the director. Yet wherever they met him, whether in the field, or the house, or the stables, there was great care on both sides to recognize one another by lifting the hat. The custom, indeed, seemed more scrupulously observed among these *farmers*, than it is in the colleges and academies of our country. And it seemed to me to exert a fine influence, and to have a strong tendency to preserve gentlemanly manners, in circumstances where men are liable to become clownish.

As we passed through the stables and noticed the fine cows, oxen, horses, and hogs, I observed that the keepers addressed M. Bella in German, and I believe some other language than the French, as it is the custom to bring men from other countries, when they introduce samples of animals from those countries, that they may be better taken care of.

In going through a stable containing a number of fine cattle, I observed one young man with water and a broom, cleaning the legs of an ox which had lain down in his leavings. The director whispered to us that that young man was the son of a wealthy banker. Indeed, the pupils all appeared as if they had not been accustomed to manual labor. Formerly, pupils were admitted from the laboring classes to attend the lectures, without residing in the institution, but they are now excluded. They now pay 750 francs, or $138, for board, and receive nothing for their labor.

This school was commenced by a society of gentlemen, to whom the buildings and farm were ceded for forty years, by Charles X. Until it passed lately into the hands of the government, it received from the government $1,100 annually. Now it is mainly sustained by the government, and the society have mostly relinquished the direction of its affairs.

Upon the whole, though I regretted that the school is closed against the poorer classes, I was much pleased with its appearance, and with that of the farm. It cannot but exert a great influence upon the agriculture of the country. Already it has sent out nearly six hundred pupils, and the present number is about eighty. For some time it suffered from the prejudices of the community against agricultural schools, and the patronage was limited. But of late the current of public opinion has changed, so that now more pupils are offered than are wanted, and the good effects of such an education are very manifest in the improved and improving condition of agriculture in the country. Such effects cannot, indeed, be reduced to numbers, as other causes conspire to produce them. But M. Bella informed me, that they were most manifest to one who had watched the progress of events. Indeed, what reasonable man can doubt, that to send forth six hundred young men, educated theoretically and practically in the true principles and practice of husbandry, must exert a powerful influence on the community ?

Mr. Colman, in his Report on European Agriculture, has given a full and detailed account of the arrangement and course of study, &c., at Grignon, which, for the most part, corresponds to the system now followed there. As his work is widely circulated, I shall not copy those details, but instead thereof, I shall give as a specimen of the instruction communicated at these schools, an abstract of the lectures delivered by some of the professors at Grand Jouan. From the samples given, the committee can judge of the manner in which all the departments of lecturing are conducted.

The subjects of study and lectures at this school are the following :—

Mathematical Sciences :—Arithmetic, Algebra, Geomety, Mechanics, Surveying, Levelling, Stereometry, (measuring solid bodies,) Linear Drawing.

Physical and Natural Sciences :—Physics, Meteorology, Mineral Chemistry, Mineralogy, Geology, Botany.

Technological Sciences :—Organic Chemistry, or Agricultural Technology, Agriculture, Arboriculture, Sylviculture, Veterinary Art, Agricultural Zoology, Equitation.

Noological Sciences :*—Rural Architecture, Forest Economy, Rural or Farm Accounts, Rural Economy, Rural Law.

* The application of the sciences to agriculture has introduced in the continental languages a new set of terms, to which we have sometimes none exactly parallel, and I have in some cases, as in that of *Noologiques*, merely anglicised the word, whose meaning the classical reader will at once understand. In this case, *Noological* is related to *Psychological*, but obviously different. In a few cases where I could not find the word in the lexicons, or found it difficult to give the exact meaning, I have transferred the word without translation, and *italicised* it.

Abstract of the course of Lectures on General Agriculture.

Agricultural Formation, (Terrain,)—1. *Soil*:—Constituent Elements, Classification of the Formations: Argillaceous, Siliceous, and peaty soils: Physical properties: Causes which modify these properties: Influence of soil on vegetation.

2. *Sub Soil*:—Sub soil active: sub soil inert: Influence of sub soil on the soil and on the life of plants.

Agricultural Geography:—Astronomic situation of France: Mountains: *Valleys, Plains, Rivers.*

Agricultural Physics:—Atmospheric Air: Caloric: Light: Darkness.

Agricultural Meteorology:—Winds: Fogs: Dew: Rain: White Frost: Frost with Ice: Snow: Hail.

Climatology:—Influence of Climate: Climate of France: Regions.

Fertilization:—Considerations preliminary: Fecundity and Fertility.

1. *Improvement*:—Clay: Rocks: Sand: Slates: Lava: *Plombage*: Irrigations: Ditching: Ploughing: Movement of the sub soil: *Colmatage.*

2. *Stimulants*:—Stimulants of Mineral Origin: Lime: Marl: Calcareous earth: Broken shells: Sea sand: the Whiting: Shell fish: Plaster: Fire Ashes: Sulphate of Iron: Salts of Potash: of Soda: of Ammonia.

Stimulants of Vegetable Origin:—Soot: Ashes: Leached Ashes.

3. *Manures*:—Animal Manures: Excrements: Urine: Pigeons' Dung: Guano: Excrement of Animals: Muscular Flesh: Blood: Fish: Fat: Oil: Woollen cloth: Horn: Horse hair: Human hair: Feathers.

Vegeto-Animal Manures:—Litter: Horse dung: of Sheep: of horned Cattle: of Swine: of Rabbits.

Animal Manures Mineralized:—Animal charcoal: Bone.

Vegetable Manures:—Green crops ploughed in. Manure and Aquatic plants: Turf: poor Vegetables: Oil Cake: Tan: Mesh: Pulpy matters: Leaves: Stubble.

Liquid Manures:—Urine of the Domestic Animals: Flemish Manures: Urine Water from *Fecularies*, (manufactories of feculant matters?)

Compound Manures:—Manure of Jauffret and Lane: Compost: Slime of Ponds: River Mud: Marine Mud.

Breaking up the Soil:—1. Work Animals: Cattle: Horses: Cows: Mules: Asses: Race: Age: Mode of tackling: Length of working: Treatment: Necessary proportion.

2. *Instruments* :—Plough with or without fore wheels : Harrow : Scarificators : Rollers : Instruments for second dressing : Weeders : *Extirpators :* Necessary proportion.

3. *Tillage* :—Theory and Practice : Soil : Temperature : Flat Tillage : Flat Tillage in rows : Flat Tillage in ridges : (?) Tillage by digging and by grubbing.

4. *Methods of moving the earth* :—Harrowing : Rolling : Second Ploughing. *Buttages.*

5. *Clearing Land* :—Heaths : Woods : Peaty lands : Clearing by the hand : by the Plough : Hoeing : Destination of the ground.

Draining :—Arable Land : Morasses : Ponds : Nature and destination of the soil.

Irrigation :—Theory and Practice : by Infiltration : Renewal of the Water : (?) *Planches Bombees.*

Quantity of water by the acre, and according to the nature of the soil. Value of the bottoms irrigated and not irrigated. Mode of working those almost irrigated. Fertility and value of the products.

Fences :—Walls : Ditches : Hedges, living or dead.

Sowing :—Theory and practice : Sowing in lines : at random : selection, renewal, cleansing, and preparation of the seeds : Burying them by the harrow : by the plough.

Method of Treatment :—Weeding : Cleaning of thistles : stripping off the leaves : *Effuillage :* Bringing into the light.

Harvesting. General Considerations.

1. *Harvesting of Fodder* :—Instruments and Machines : Mowing : Hay making : Grindstones.

2. *Harvesting of Grain* :—Instruments and Machines : Mowing : Reaping : Threshing : *Liage.*

3. *Harvesting of Roots* :—Pulling up by the hand : by the plough : Uncovering : Cleaning.

Selection of the methods of preparing the Soils :—According to atmospheric circumstances : Nature of the Soil : its condition : its destination.

Distribution of Labor by Rotation : (?)—Normal conditions : Exceptional conditions.

Rural Architecture.

Materials :—Siliceous, calcareous and argillaceous rocks : Fat, meagre, and hydraulic Lime : Sands : Mortar : Cements : Puzzolana : Plaster : Wood : Iron : Paving Brick : Roofing Slate : Tiles : Lead : Zinc : Leather : Ropes.

Works :—Foundations : Terracing : Properties of Earths.

Masonry :—Foundation Walls : High Walls : for support : for enclosure : Plastering : Pise.

Carpentry :—*Assemblages* : *Combles* : *Pans de bois* : Partitions : Staircases.

Joiners' Work :—Floors : Gates : Windows : Shutters.

Iron Work :—Large Iron : Ironing the Buildings.

Roofing :—Tiles : Slate : Thatch : Zinc : Bitumen.

Painting and Glazing :—Oil Painting : Distemper Paintings : Badidgeon, (coloring) Window glass.

Paving and Bricking.

Estimate of the Works :—Masonry : Carpentry.

Specification :—Form of the works.

Edifices :—Stable : Cow house : Sheep fold : Hog pen : Hen house : Pigeon house : Silk worm nursery.

Animal products :—Dairy : Cheese house.

Vegetable products :—Barns : Granaries : Wine cellars : Cellars : Corn pits : Ovens.

Agricultural Manufactures :—*Feculary* : Distillery : Sugar manufactory.

Reservoirs :—Watering places : Wash house : Wells : Cisterns : Ditches for urine : Ponds.

Dwelling house :—Form and Proportion.

Irrigations :—Dams : Taking out the Water : Sluices : Canals : Weirs : Slopes.

Drainage :—Damming up : Trenching : Cesspool : Machines for drainage.

Routs :—Soil : Slope : Outline : Levelling : Materials : Support : Bridges : Estimate of Excavation and Embankment.

Group of Edifices composing a Farming Establishment :—Relation to the fertility of the soil and the culture and extent of the farm.

National Agronomic Institute.

The general character of this institution has already been given, in the law of the 3d of October, 1848. By that law, it is located at Versailles. Three farms, a garden, and a forest, embracing about 3,452 acres, and a portion of the public buildings, have been devoted to it. This is one of the most signal triumphs of republicanism in France, that so large a portion of that magnificent seat of royalty, should be devoted to an agricultural institution. It is not yet, indeed, got into full operation, but the administration is in full possession of the buildings in which it is to be placed. A collection of the skeletons of mammiferous animals, and of parasitic insects, with colored figures, a series of photographic plates of animals for the farm, &c., are already on the ground, and wait only for the fitting up of the halls for their reception, as does also a library. Some of the professors and other officers have been appointed, and the school will soon be in operation; perhaps it is before this time.

I cannot perceive that this Institute differs much in its organization from that of the regional schools, except that it is on a more extended scale, and the course of study more elevated. It presents itself in a three fold aspect:

First, as having a Faculty of the agronomic sciences.

Secondly, as a superior normal school for agriculture.

Thirdly, as a higher institute for agricultural education, open to the administrators and proprietors who have turned their attention to agriculture.

To meet the wants of this latter class especially, a large farm is connected with the school. Here will be performed, at the expense of the state, all the experiments necessary to the progress of agronomic science, and to verify practically all the innovations and improvements proposed by others, before they are recommended to the public. This is an object of great importance, and should enter into any plan for a school in the United States.

The theoretical and practical parts of this institute are really distinct, but they are placed under the general government of one director.

The professorships are nine, as follows:

One chair of rural economy and legislation.

One of agriculture.

One of zootechny, or the economy of animals.

One of sylviculture.

One of rural engineering, embracing levelling, irrigation, construction of roads, rural architecture, and mechanics applied to agricultural instruments.

The above professorships belong to practical agriculture. The others belong to the theory of the subject.

One of terrestrial physics and meteorology.

One of chemistry applied to agriculture.

One of botany, and vegetable physiology.

One of applied zoology.

Here, as in the lower schools, a number of tutors is appointed equal to the number of professors.

In addition to the director, professors, and tutors, the following officers will be appointed:

A prefect of studies.

A curator of the collections.

A librarian.

An overseer of studies.

To these will be added a corps of head men to oversee and manage the affairs of the farm. These will in part be called from the farm schools. For example, the institute will need twenty-one herdsmen, twenty-one grooms, twenty-one shepherds, and fifteen gardeners.

The French minister adds, " The end of the institute at Versailles is not merely to afford agricultural instruction, but to open the way for studious men, who wish to direct their labors towards the application of science to rural industry. This is the first attempt of the kind that has been made. Industry has enriched the learned men who have explored the domain of the physical sciences and of chemistry for this object. But if agriculture has given reputation to any, it has not procured for any one a position which would enable him to make that the centre of his studies. Hence scientific teaching everywhere has become industrial, (?) because there only can it be repaid for its services. The institute at Versailles is intended to change this state of things by offering as a prize of laudable ambition, to those who direct their researches to agriculture, a certain number of chairs, before which an immense field opens."

Much more might be added respecting this school, but I have probably given the most important facts.

SWITZERLAND.

I am not aware that any agricultural schools exist at this moment in Switzerland, but one has been in operation there which has excited a good deal of interest.

The Hofwyl School.

This institution, established by M. Fellenberg in 1799, has been often described, and has excited deep interest in Europe and America. But it has some time since ceased its operations, at least in almost all departments. I visited the place, however, (nine miles from Berne, in a beautiful country, and over a beautiful road,) not to see the agricultural school, but to learn why it had been discontinued; a point, I thought, of much importance to those who are deliberating whether to establish one. I found a fine farm and extensive buildings, large enough for a university; and I could easily see how it was, that after the death of Fellenberg, the father, in 1844, his two sons found their task so severe that they could not sustain it. The intelligent gentleman who showed me over the establishment, (connected, I believe, with the family of Fellenberg, but his name I have forgotten,) told me that the school was given up, not for want of patronage, for that was abundant, but because the sons of Fellenberg lost their health, through too severe labor. From filial piety they even held on longer than prudence allowed. The fact is, the establishment in all its parts was little short of a university, with agriculture as its basis. It comprehended the following departments:

1. A farm, intended as a model.
2. An experimental farm.
3. A manufactory of agricultural instruments.
4. A workshop, employed in the improvement of agricultural mechanism.
5. A school of industry for the poor.
6. A seminary for children of the higher class.
7. An academy of theoretical and practical agriculture.
8. A school for the instruction of masters.

It was an attempt to bring together the sons of princes and nobles, along with those of the peasantry and orphans. The school of the former was very expensive, and intended to furnish the means of sustaining the latter almost gratuitously. To the former were attached not less than thirty professors, after Fellenberg, at the suggestion of the Emperor of Russia, had so enlarged it as to accommodate one hundred pupils; and to these were added masters in music, drawing, and all other branches of a refined and liberal education. How could such a vast system be sustained by one man, or even two, upon a farm in the interior of Switzerland? especially when it kept up the distinctions between the rich and the poor so fully, as to excite the warm approbation of the autocrat of all the Russias. I doubt not that when Fellenberg yielded his own judgment to such an adviser, and greatly enlarged the aristocratic branch of his school, he struck a death blow to its permanence. He was doubtless actuated by benevolent motives, and his school has done good, by exciting public attention to the system which he endeavored in vain practically to realize. But he attempted to reconcile and hold together by a principle of love, what can be sustained only by the bayonet, viz., hereditary distinctions of caste and enlightened poverty. Education and religion do indeed create distinctions among men, but they are founded upon merit and virtue, not upon the accidental and unimportant circumstances of birth and pedigree. The former are essential to the existence of society, but the latter will be less and less regarded, as education and true religion take the place of soldiers and church establishments.

The farm at Hofwyl appears to be a good one, and there are some students there still, who receive some instruction, and labor on the land; there is also a small orphan school. The collections in natural history are not large; they have numerous agricultural instruments, and a manufactory of the same, though on a small scale at present.

After seeing the remains of this establishment, I rather wondered that it had not stopped sooner. While it continued, I was told that the farmers of the surrounding country took much interest in it. Its failure does not seem to me to furnish a ground of discouragement to others who would establish agricultural schools, especially, in a country where the sons of the rich do not think themselves degraded by forming a part of the same school with those of the poor who have equal talents; and of this character to a great extent are all the schools of our country.

ITALY.

The precise state of agricultural education in Italy it is difficult to ascertain, since political changes affect it. But in 1842, a school of the superior class was established in connection with the university of Pisa, called an agronomic and pastoral institution; the term *pastoral*, embracing the entire management of animals. It is placed at the gates of the city, where there is a garden, a veterinary establishment, and a shop for the construction of agricultural implements. One professor is attached exclusively to the institute; the others belong to the university, and lecture in the institute. The first year the instruction embraces geometry, algebra, general physics, and botany; the second year, descriptive geometry, surveying, chemistry, and agronomy; the third year, geology, applied physics, rural architecture, agronomy, and the care of animals, (la pastorale.) Some of the lectures are given at the university, and some at the institute. Students aid in the practical labors of the establishment a certain number of hours during the second and third years. They are examined each year upon the ground gone over, and if accepted, receive, at the end of the third year, a license for the agricultural sciences. Students are admitted, also, for a term of three months, or more, who do not finish the regular course.

Another school of inferior grade did exist at Meleto, founded in 1834, but I have no information concerning it.

BELGIUM.

It is only during the past year that the government of Belgium have established agricultural schools. Nine of these are now in operation, of which three are intermediary, two inferior, one special, and three are connected with the communal colleges,—a sort of gymnasium. The course of instruction occupies three years, and in the intermediary schools embraces geometry, surveying, levelling, drawing, mechanics, physics, mineralogy, geology, botany, chemistry, agricultural technology, horticulture, the economy of forests, agricultural zoology, the veterinary art, hygiene, rural law, rural architecture, and agriculture. In the inferior schools it embraces elementary mathematics, surveying, levelling, drawing of plans, elements of the phyical and natural sciences, and agricultural book keeping.

A committee, appointed by the government of the state, visits the schools at least once every three months. The students are of two classes ; those paying board, and those supported by the state.

The officers of these schools, (*le personnel*,) are a director, from two to five professors, an instructor in gardening, an overseer of the practical operations, and as many workmen as are necessary. Four of the schools are independent ; the rest are connected with other institutions ; three with communal colleges, and two with industrial schools. The number of scholarships and the amount of tuition, may be seen on the annexed schedule, with a few other facts.

PRUSSIA.

The agricultural schools of Prussia present us with all the varieties that have been enumerated, viz.: the superior, intermediary, inferior, special, and connected with the universities.

The superior schools, called Royal Institutes or Academies of Agriculture, are three in number, at Moeglin, Eldena, and Poppelsdorf.

That at Moeglen was established forty-four years ago, upon a farm of 2,480 acres. It admits only twenty pupils, to board with the director, at $240 per annum, yet it has sent out 511, who have finished a four years' course. Besides these, the school admits pupils temporarily at $7 per week.

The instruction embraces agriculture, rural economy, the raising of cattle, the mathematical, physical, chemical, and natural sciences, applied to agriculture, rural industry, and the veterinary art. It also gives more special instruction in farm accounts, in the raising of animals with fine wool, the irrigation of meadows, and the cultivation of potatoes on a large scale.

The institution at Eldena in Pomerania, is situated on an estate belonging to the university of Greisswalde, which is near. It has been in operation about twenty years ; and of its eighty pupils, about seventy are intending to devote themselves to agriculture, and ten to the higher departments of the government, where they will need a knowledge of agriculture. The course is two years, and the cost of instruction, not including board, is about $90. The age of those admitted must be at least seventeen years, and they must produce testimonials of good conduct, and of having pursued classical studies.

The nine professors at Eldena give ten courses of lectures on the following subjects:—1. *Political Economy*:—Finances: Rural Police: Constitutional Law in Prussia: Governmental organization: Politico-economic discussions. 2. *History and Statistics of Agriculture*:—Agriculture in general: Agriculture special: Cultivation of meadows: Zootechny in general: Raising of sheep: Raising of horned animals: Rural economy: Systems of culture: Valuation of rural estates: Agricultural book-keeping, theoretic and practical. 3. *Sylviculture* in general, (culture of groves.) 4. *Horticulture*:—Culture of garden vegetables: Culture of fruit trees: Arboriculture, (culture of trees and shrubs for timber, &c.) 5. *Raising of Horses*:—Anatomy and physiology of domestic animals: Veterinary medicine: Hygiene. 6. *Chemistry*:—Experimental and Agricultural, organic and inorganic chemistry, (exercises in the laboratory,) Physics and Meteorology: Technology with practical demonstration in the distillery: Brewery: Tile kiln, and dairy: Excursion to the saline of Greisswalde: to the beet sugar manufactory of Stralsund: Manufactory of instruments and mills. 7. *Anatomy, Physiology, and Geology of Plants*:—Botany, general and applied to agriculture: Horticulture and sylviculture: Zoology, general and applied to agriculture: Excursions. 8. *Arithmetic and Mathematics*:—Surveying: Levelling: General and applied mechanics. 9. *Drawing*:—Rural architecture: Practical valuation of constructions. 10. *Rural Law.*

The Superior Institute at Regenwalde, in Pomerania also, is established on essentially the same plan as that at Eldena. It has four professors, who give eleven courses of lectures, and the pupils pay about $221 per annum.

The institute at Popplesdorf, near Bonn, and connected with its university, is modelled on a similar plan. But as I had an opportunity to examine it, I will add a few words respecting its character.

Popplesdorf is only a mile and a half from Bonn, through a most beautiful shaded walk. The splendid collections in natural history, belonging to the university, are at Popplesdorf, occupying the rooms of an ancient palace. Close by, is the model farm of the agricultural institute. There are two professors (Drs. Schweitzer and Hartstein,) connected directly with the institute. They, however, have liberty, if they can obtain classes, of lecturing in the university. The more scientific lectures, however, are given by professors in the university, and the library and collections of the same are accessible to the agricultural students. I was indebted to Prof. Reisin of the university, for introducing me to Dr. Hartstein, and for acting as interpreter between us.

The studies taught at this school by its nine professors, who give ten courses, are so similar to those at Eldena, that I need not repeat the list. In order to admission, the students must be entered on the register of the university, in the faculty of philosophy, which, in Germany, embraces all that is not theology, law, or medicine. They must also produce certificates of good conduct, and of having studied the classics. A few students of the other faculties in the university attend the agricultural lectures, but not many. The course is two years; the annual expense, $210; the present number of pupils is forty-seven, and the school has been four years in operation. Most of the pupils are sons of wealthy farmers, and they become, after leaving, farmers themselves, or directors of farms.

The farm consists of about seventy acres of good land. I saw on it a great variety of crops, several of which were raised merely for experiment. The cultivation appeared to be good, though hardly equal to what I had seen on the model farms in Ireland. The barns and cattle as well as agricultural implements appeared to be of a superior character. Although the school has been in operation only four years, I was assured that it exerted an excellent influence upon the agricultural interests of the region. It is, however, far less liberally endowed than the school at Eldena; yet it has great advantages in the libraries and collections of the university.

I ought not to forget to notice my obligations to Chevalier Bunsen, who happened to be at Bonn at the same time as myself, and who gave me a letter of introduction to some of the professors.

Another superior school, with eight professors, and a farm of 2,312 acres, exists at Proskau, whose organization presents nothing peculiar.

The intermediary schools prepare the pupils for the higher institutes. The instruction is exclusively practical, save some in the winter on the experience of the summer. The price of board is $69 per annum.

In the inferior schools the pupils take the place of hired servants. Accordingly they pay little or nothing, and sometimes receive wages. They devote themselves to manual labor, to operations with cattle, and with horses. The number of schools is twenty-three, of which twelve are for teaching agriculture, two for the cultivation of meadows, irrigations, &c., one for training shepherds, and eight for the raising, dressing, and working of flax.

The whole number of schools in Prussia, then, is thirty-two; of which five are superior; two intermediary; twelve inferior; thirteen special. Of the superior, two are connected more or less with universities.

AUSTRIA.

Austria possesses some of the oldest agricultural schools in Europe. That at Krummau, in Bohemia, was founded in 1799, by prince Schwartzenberg, on a domain containing nearly 300,000 acres, in the same year as the Hofwyl school. The means of instruction are on a large scale, and the collections are very ample. They embrace a collection of minerals; one of philosophical and chemical instruments; one of models of agricultural and other instruments and machines; one of the phalenas (insects,) of the country; an astronomical observatory; a botanic garden; a conservatory; an herbarium; and a pomological cabinet. There are four farms cultivated according to different systems. The instruction is gratuitous, and the professors are charged to join practice to theory, as they have a good opportunity of doing, upon the ample farms. In Styria, at Graetz, is another superior school founded in 1809, in which nine professors give eleven courses in botany, zoology, mineralogy, geology, mathematics, chemistry, physics, mechanics, agriculture, sylviculture, and the working of mines. Here, also, are collections in natural history, and in agriculture, and a rich library, as well as a large botanic garden. A small model farm is attached, as well as a model *magnanery*, (establishment for silk worms.) The instruction is gratuitous, and there are ten scholarships.

Another school of the industrial class, but of superior grade, was founded in Prague, in 1803, and has six professors, who give eight courses of lectures, viz.: in agriculture; chemistry, general and applied; applied mathematics; architecture; mechanics; hydraulics; drawing, and technology.

A fourth superior school was founded in 1848, at Cracow, in Austrian Poland, by Count Adam Potocky, with a capital of $4,000, and an annual income of $1,600.

Attached to the university of Lemberg, the capital of Gallicia, is a professorship of agriculture, and another of sylviculture. But I have no particulars concerning the instruction.

Of the inferior schools, that of Trutsch was founded by the countess Dietrischstein, for giving instruction to the young peasants in theoretical, but mostly, practical agriculture. That at Koeingsheal has sixty boarding scholars, besides many others.

The special schools, amounting to twenty-five, are mostly for the culture of flax. The course lasts one year, but some stay longer. There is an annual examination, and four prizes are distributed. Many thousands have received instruction in these schools.

Such schools as these may at first view seem of no consequence in this country. But if the recent improvements in the mode of converting flax into cloth, with wool, prove as valuable as the English newspapers represent, (the London Morning Chronicle,) such schools, or at least the instruction communicated in them, may become of great importance to Massachusetts, since, in that case, flax would probably be substituted for cotton.

The whole number of schools in Austria is thirty-three, of which four are superior, three inferior, twenty-five special, and one connected with a university.

WURTEMBERG.

The agricultural schools of this German state have long been known for their superior character. They are sustained and directed entirely by the state, while those of Prussia are carried on by individuals, aided by the state. M. Royer considers this difference a sufficient reason for the superior character of the Wurtemberg schools.

The Royal Institute of agronomy and forests, at Hohenheim, is the most extensive and best known of these schools, the only one, in fact, of the superior class. It was founded by king William, in 1818, on the royal domain of some 825 acres, not including 3,000 acres of forest. This property is given up entirely to the management of the school, which renders an annual account of the expenses and receipts, and any deficiency is supplied by the state.

The *personnel* consists of one director, six professors, and four functionaries charged with various labors, besides two tutors, who hear lessons in the school, although residing abroad.

The instruction is embraced in forty courses, divided into three groups:—1. Agricultural matters. 2. Forest matters. 3. Auxiliary sciences. A brief statement will give an idea of the instruction.

1*st course.*—Agriculture, properly so called. 1. Of climate; 2. Of soil; 3. Of manures; 4. Of tools and instruments for tillage; 5. Of clearing up the ground; 6. Of meadows and pastures; 7. Of agriculture in general; *a.* Of ploughing and other tillage; *b.* Of seed plots; *c.* Of tillage following grain crops; *d.* Of crops; *e.* Of threshing and the preservation of grain; 8. Of special agriculture. All cultivated plants are treated of particularly.

2*d course.*—Viticulture; 1. Culture of the vine; 2. Wine making.

3*d course.*—Culture of fruit trees.

4*th course.*—Raising of cattle; The races; The crossing; The young.

5*th course.*—The raising of the horse; Natural history of the horse; Different methods of raising; Choice of animals for reproduction; Treatment of mares; Treatment of colts.

6*th course.*—Sericulture, (silk culture,) Culture of the mulberry; Management of the silk worms.

7*th course.*—Rural industry; *In winter:* Manufacture of beet sugar; Of liquid manure? Of malt, beer and brandy. *In summer:* Manufacture of liquors; Of vinegar; Of cider; of lime and of tiles.

8*th course:*—1. *Rural economy;* 3. Valuation of rural estates; *a.* General circumstances of the country; *b.* Of farms in general; *c.* Of different parts of a farm; *d.* Of the means of maintaining the fertility of the soil; *e.* Of systems of culture; *f.* Of labor and the internal organization of a farm; *g.* Relation between the number of beasts and the extent of the land worked; *h.* Of the capital of the undertaker; *i.* Of different modes of working a farm.

9*th course:*—Agricultural book keeping; In general; Practising book keeping.

10*th course:*—On the operations at Hohenheim, &c.

2.—Forest Matters.

1*st course:*—Encyclopedia of the science of forests.

2*d course:*—Forest botany.

3*d course:*—Sylviculture.

4*th course:*—Forest technology; *a.* Working of wood; *b.* Accessory products of the forests; *c.* Of forest industry.

5*th course:*—Preservation and defence of forests; *a.* Against the attacks of man; *b.* Do. of animals; *c.* Do. against natural accidents.

6*th course:*—Hunting.

7*th course:*—Forest work; *a.* Of the inventory; Surveying and description of a forest; *b.* Of the statistics of do.; *c.* Of the systems of cutting out the wood; *d.* Of the valuation of forests.

8*th course:*—Forest economy; *a.* Political economy applied to forests; *b.* Forest labors; *c.* Forest police; *d.* Of the government of the state forests.

9*th course:*—Forest legislation and rules of service for forest agents in Wurtemberg.

10*th course:*—Practical part of forest concerns.

3.—Auxiliary Sciences.

1st course:—Higher arithmetic.

2d course:—Algebra.

3d course:—Planimetry.

4th course:—Stereometry.

5th course:—Trigonometry.

6th course:—Applied geometry.

7th course:—Mathematics applied to forests; 1. Of the culture of trees and of the entire forest; 2. Of the increase of trees; 3. Of the valuation in money of forests.

8th course:—Physics.

9th course:—Mechanics.

10th course:—Chemistry.

11th course:—Oryctognosy.

12th course:—Geognosy.

13th course:—Vegetable botany and physiology.

14th course:—Special and rural botany.

15th course:—Zoology.

16th course:—Veterinary medicine; 1. Natural history of our domestic animals; 2. Anatomy of do.; 3. Animal physiology; 4. Care to be taken of animals; 5. Of the medicines proper for slight diseases; 6. Description of diseases; pathology and therapeutics; 7. Veterinary surgery; 8. Internal diseases of animals, and murrains.

17th course:—Of forest law.

18th course:—Rural constructions.

19th course:—Of preparing plans.

20th course:—Drawing of machines.

To illustrate these courses of instruction, the means seem to be very ample at Hohenheim. They are as follows:—

The operations on a large farm annexed to the Institute.

A forest of 5000 acres.

A botanic garden.

A library open twice a week.

A geological collection.

A mineralogical do.

A botanical do.

A collection of woods, seeds, and resins from the forest.

A collection in comparative anatomy.

Do. of specimens of wool.

Do. of agricultural products.

Do. of models of instruments for tillage.

Instruments for surveying.

Do. for physical science.

Do. for the chemical laboratory.

The students board where they please, at a price from $24 to $120 per annum, but lodge at the institute.

The number of students in 1849 was about 100, but it had been 140 for many years. No less than 1650 finished their education at this seminary within thirty-one years. And how is it possible that so many, having gone through such a thorough system of instruction, should not exert a powerful influence upon agriculture throughout the community?

There is also at Hohenheim, a practical school of agriculture for orphans and poorer children of peasants, who receive instruction gratuitously, and are paid for their labor. Also a school of horticulture, whose special object is the cultivation of fruit trees, of garden vegetables, and ornamental trees. The pupils do all the work, yet receive five lessons in horticulture in the week, and one or two on agriculture each day. There is also, at the same place, a school for meadows and irrigations,—theoretical in the winter and practical in summer. The pupils receive wages. A school likewise exists in the same place, for the cultivation of fruit trees, for the preparation of flax, and the manufactory of agricultural instruments. Schools of agriculture have been also opened in a few other places in the state, in which the course of instruction appears to be rather extensive.

The whole number of schools is seven; one superior, two intermediary, one inferior, and three special. Probably nowhere in Europe shall we find a better model for such schools than at Hohenheim.

BAVARIA.

Bavaria has one school of the superior class at Scheissheim, near Munich. It was founded in 1822, on a national domain of 6,415 acres, on which from 350 to 500 head of cattle are supported. There are, in fact, two schools; one superior, to teach theory, and one practical, to apply, test, and elucidate theory. The instruction in the superior school embraces eleven courses, so similar to those at Hohenheim, that I need not repeat them. The lower school is intended to be preparatory for admission to the superior. The examination for such admission requires the candidate to be sixteen years of age, to have gone over the studies of the lower school, and to understand "a little of Latin, to be able to comprehend the terminology of agriculture."

Not less than thirty-two inferior schools have been established in as many towns of the kingdom, but they are not in all cases in a flourishing state, and need no detailed description. They are, in fact, for the most part, industrial schools, in which agriculture is taught.

The school at Litchenhof, near Nuremberg, of a mixed or intermediary class, deserves notice. It was founded in 1833, by Dr. Wedenkeller, and sustained for a time by men of wealth. His success determined the government to create scholarships. It comprehends, in fact, four different schools in a three years' course.

1. A preparatory school,—agricultural and industrial.
2. A school of horticulture.
3. A school for head servants.
4. A school for herdsmen and shepherds.

The subjects taught the first year are religion, German language, geography, arithmetic, zoology, agriculture, agricultural arts, drawing, and practical agricultural exercises.

The second year;—The same subjects pursued further; with geometry, botany, mineralogy, history.

The third year;—Religion, German language, compositions, agricultural chemistry, farm accounts, agriculture, rural architecture, machines, anatomy of animals, veterinary medicine, drawing, riding and fencing, practical exercises.

There are also facilities for learning Latin and French.

This school should be placed as high at least as the intermediary class; and then we shall have in Bavaria, one superior school, one intermediary, thirty-two inferior, and one special; in all, thirty-five.

SAXONY.

This kingdom has one superior school, three intermediary, and one special; five in all.

The superior school is at Tharand, and was established in 1811, by H. Cotta, simply as a school for forests, but afterwards, in 1829, agriculture was added. The domain contains 7,355 acres, and there are nine professors; two of them for the forests. The two years' course of instruction scarcely differs from that of the better schools already described.

From 1816 to 1841, no less than 826 pupils left this school, of whom 701 devoted themselves to the forests, and 125 to agriculture. But the latter branch was not introduced till 1829. Since 1841, the mean number of pupils has been about 80, and the whole number since the foundation of the school has been 1,100. It has a library, collections of natural history, and a garden. Every year the school makes extensive excursions, or takes a scientific voyage.

The course is two years long. Charge for tuition varies from $35 to $52 annually, foreign students being charged the highest.

The intermediary school at Dresden, is limited to ten or twelve pupils. The course of instruction is ample, and similar to that in other schools. More facilities seem to be afforded here for the study of languages than is usual. Greek and Latin are among the number, with other modern languages besides the German.

A similar school exists at Broesa near Bautzen. In this school the summer is devoted to the practice and the winter to the study of agriculture. Here also, as in some other German schools, agricultural discussions are conducted by the pupils.

Another similar establishment exists at Schoenfeld, near Dresden. Here instruction is embraced in six courses ; the same as already noticed, with the addition of the theory and practice of working iron. Excursions are made by the pupils to well conducted farms. There are scholarships for one in five of the pupils. The charge for instruction varies from $17 to $70.

A school of the special class for orphan children exists at Groshennersdorf, on a farm of fifty-eight acres. The special instruction is based on the labors of the farm, of which three-quarters are performed by the spade, and one quarter by the plough. To the cultivation of fruit trees they add that of the mulberry, and raise silk worms and bees. The pupils labor about seven hours per day. The ordinary instruction embraces religion, reading, writing, arithmetic, German grammar, a little of geography, singing and gymnastic exercises. They enter at eight years and leave at fourteen.

BRUNSWICK.

Brunswick has only two schools, one superior, the other inferior. The object of the former is to give that scientific instruction which forms the basis of various branches of industry. Practice is not employed except to make theory understood.

This school, located in Brunswick, has thirteen professors, who give thirteen courses, as follows :—1. Physics and general chemistry ; 2. Mathematics, pure ; geometry and linear drawing ; 3. Mineralogy and zoology ; 4 Architecture ; 5. Transcendental mathematics ; 6. Rural economy ; 7. Applied chemistry ; 8. Mechanical technology ; 9 Natural history ; 10. Raising of horses, pathology, and therapeutics of domestic animals ; 11. Veterinary medicine ; 12. Forest science ; 13. Architecture. (There are two professors of architecture.) The means of instruction are very ample ; consisting in cabinets, laboratories, and collections, for every kind of instruction ; also a botanic garden, a ducal museum, stables, &c. Excursions are made to the various agricultural manufactories, and to the forests ; and experiments on a large scale are carried on upon two adjoining farms.

The inferior school is a private establishment for the sons of peasants, with a course of two years' instruction, having three professors.

MECKLENBURG SCHWERIN.

The only agricultural establishment in this state is a private one, at Carlshof, belonging to the intermediary class. Its object is to give sufficient instruction to future cultivators and farmers. In the list of studies, I notice no peculiarity, except one item, which teaches the values of foreign moneys, the importance of which, any one can appreciate who has travelled in Germany.

SCHLESWIG HOLSTEIN.

This state reckons four agricultural schools ; two intermediary, and two inferior. They are founded and sustained by individuals, or societies, and not by government.

The agronomic institute of Toestrup, has about twenty pupils. A field for experiments is provided ; also a nursery of fruit trees ; a forest nursery ; collection of model instruments, and a library.

Of the inferior schools, that at Rodding was founded by a society in Copenhagen, which pays $240 annually to a professor of agriculture. That at Jevenstaed numbers about forty scholars.

PRINCIPALITY OF ANHALT.

This principality contains two schools, one superior, and one intermediary. The former, at Coswig, has ten professors, and the course is one year. The instruction is divided into four parts :—1. Practical instruction in the manufactures of the institute. 2. Regular courses on agricultural technology, and the auxiliary sciences, such as physics, chemistry, political economy, and the laws relating to agricultural matters.

The course on technology embraces the manufacture of alcohol, of liquors, of vinegar, of leaven, of syrup, of sugar, of oil, of beer, of butter, of cheese, of lime, of tiles, &c.

3. Discussions in the presence of the professors.

4. Manual labors.

The intermediary school at Cocksted was founded in 1837, for irrigations, and agriculture. Its course lasts only a year ; that part in summer being devoted to irrigations, and that in winter to agriculture.

Of the thirty-two pupils in this school from 1839 to 1847, seventeen were preparing to become agriculturists ; six civil engineers ; six devoted to the science of forests, and three functionaries. Eight of these were Danes, five Swedes, and three Russians; the other half natives of the country.

GRAND DUCHY OF HESSE.

Of the two schools in this state, the one situated at Darmstadt is intermediary, and the other, at Arnsburg, is for practical agriculture, but placed in the same rank. There is nothing in their organization that needs particular notice in this connection.

GRAND DUCHY OF WEIMAR.

The only school in this duchy is of the superior class, situated at Jena, and connected with the university in that place. Its objects are, 1. To perfect young agriculturists in the theory and practice of agriculture. 2. To afford to those already practical agriculturists, the means of following the course at the university. 3. To prepare functionaries for the different departments of the government. The pupils rank with those of the university.

The lectures in the university, attended by the agricultural scholars, embrace political economy, history, philosophy, mathematics, natural history and law. The instruction in the school proper embraces a great variety of subjects, among which I notice none not already named, unless it be the art of shoeing horses, in connection with veterinary medicine. To this instruction are added discussions, compositions on agricultural subjects, analytical exercises in the laboratory, visits to the botanic garden, and the collections, and to farms well managed. The yearly tuition is **$30**, and the whole expense **$210**.

DUCHY OF NASSAU.

The Agronomic Institute at Geissberg, one and a half miles from Wiesbaden, is the only agricultural school in this duchy, and is of the superior class, though its organization is somewhat peculiar. It was founded in **1835**, by the state, and under the impression that the theory and practice of agriculture could not be well united in the same school. The pupils remain at the institute only during the winter, and in the summer they are at home, or with good farmers. There is, however, a farm of about seventy-five acres attached to the school.

I visited this establishment, and found it delightfully situated, on an estate of the duke of Nassau, a little north of the village of Wiesbaden. The buildings are upon a hill, well cultivated, which rises a few hundred feet above Wiesbaden, and overlooks the broad valley of the Rhine; Mayence lying in the distance; and other towns are visible.

The buildings are in fine style, and the rooms devoted to the collections, of a superior character. In them I found collections of rocks, birds, quadrupeds, seeds, grains, and grasses; a pomological cabinet in wax, models of agricultural instruments, and a good library. The course lasts two years, and the instruction is given by lectures and recitations. The pupils number at present about fifty. Government pays about $18 for each pupil, and there are six scholarships, of $10 each, for the most regular and laborious. Natives of the country pay nothing for tuition; foreigners pay about $18. Those who board out of the establishment pay from $70 to $140 for board.

For the first winter, the course of instruction in this school embraces the German language, arithmetic, mineralogy, botany, physics, general agriculture, general zootechny, culture of meadows, rural architecture, and veterinary medicine.

For the second winter, compositions, zoology, physics, farm accounts, special agriculture, special zootechny, horticulture, technology, and veterinary medicine.

All the agricultural schools which I have visited in Europe, like that at Geissberg, have a beautiful location; and I think this a point not to be overlooked in fixing upon a place for such a school, for scenery has a great influence upon the youthful mind and heart.

ELECTORATE OF HESSE.

The only establishment for agriculture in this electorate, is the Institute at Beberbeck, founded in 1846, on a domain belonging to the elector, containing, besides pasturage, 620 acres. Its course of instruction is ample, as are its collections. But there is little that is peculiar, or that has not already been described. The school belongs to the superior class.

GRAND DUCHY OF BADEN.

This duchy has only one school, and that of the intermediary class, situated at Hochburg, founded in 1848, on the national domain. Its course continues three years. Its course of instruction is ample, but need not be given here. The pupils have twelve lessons a week in the winter, and seventeen in summer. As to practical instruction, the first year is confined to ordinary manual labors, the second to the care of animals, and the third to spans of horses in harness, and to all sorts of labors.

The pupils receive wages; the first year, $28; the second year, $37; and the third year, $46; but the cost of board is $70. Instruction is gratuitous, and the state pays to each pupil, a gratuity of $16.

DUCHY OF SAXE MEININGEN.

Only one school, and that of the inferior class, exists in this duchy, founded in 1847, at Frauenbreitungen. I cannot find that it has anything peculiar to it, not already described. Its object is to prepare practical cultivators, not neglecting the theory.

RUSSIA.

Within a few years past, the government of Russia has been much awake to the subject of agricultural schools. Not less than sixty-eight of these have been established, two of which are superior, ten intermediary, fifty-one inferior, four special, and one connected with a university.

The Agronomic Institute of St. Petersburg was founded in 1833, on a territory of 330 acres. Its organization cost 700,000 roubles, ($525,000,) and its annual support, 100,000 roubles, ($75,000.)*

Its objects are :—1. To give theoretical and practical agricultural instruction to the children of the imperial serfs. 2. To try the different systems of cultivation practised abroad, to ascertain which of them will answer for Russia. 3. To form directors for the model farms. The course lasts five years, and embraces religion, sacred history, the liturgy ; discussions, moral and instructive ; mathematics, pure and applied ; surveying and agriculture. The number of pupils is 250, who are allowed to enter at fifteen. In giving instruction, the method of Lancaster is adopted, and it is more practical than theoretical.

The *personnel* consists of a director, a sub director, a priest, a physician, a steward, professors, and overseers. The labor is performed by the pupils, even such as shoemaking, locksmithing, joinery, and cabinet work. There is a library, a brewery, and particularly, numerous workshops.

The Imperial Institute at Gorgigoretz embraces, in fact, a superior, intermediary, and inferior school. In the first, there are 200 pupils, sons of the peasants, supported by the state, at an expense of $38 each, annually. In the second, are thirty-five scholars, who pay $66 each, for board, clothing, &c. In the third, the course is three years, and the pupils, eighty-five in number, come from the middle classes of society, and they pay $84 for better accommodations.

To take charge of this superior school, there is one director, a professor of agriculture, and six adjunct professors, of whom two are

* Reckoning the rouble at seventy-five cents.

for agriculture, rural economy, and sylviculture; two for natural sciences; and two for public and administrative law, statistics, and literature; also eight professors for teaching religion, the Russian and German languages, mathematics, history, geography, architecture, geodesy, drawing and writing.

The other officers of this higher school are an inspector, and three adjuncts, one veterinary physician, one steward, one cashier, one chief secretary, and one accountant.

Another school of the superior class, is the Agronomic Institute of Moscow, founded in 1843, whose term of study is five years, whose farm is 660 acres, and whose course of study is ample. The pupils pay about $100 per annum. The professor of agriculture is connected with the university of Moscow, but what other relations the school has to the university I cannot ascertain.

Two schools, one of the intermediary and one of the inferior class, have been established at Marjino, by the countess Straganow. To the first, she gave $11,000, and it had one hundred and twenty-five pupils in 1844, of whom sixty-four were intended for agriculture, thirty-six for the mines, seventeen for forest economy, and fourteen for surveying. In this school are twenty professors, with a salary each of $825. The student pays $138 annually.

Since 1845, not less than fifty inferior schools have been established in connection with model farms. Some of them have a four years' course, on farms varying from 1,375 to 2,220 acres, with from 100 to 200 pupils. They are really of a higher grade than the farm schools of other countries.

Of the four special schools, one is for horticulture, one for silks, one for bees, and one for flax. But further details respecting such schools can be of little use.

SWEDEN.

Professor Johnstone, in his Agricultural Address at Syracuse, in Sept. 1849, speaks of agricultural schools and model farms as existing in Sweden, and says that one of these establishments "each province is expected in a few years to possess." But he has given no details, nor are any within my reach.

SUMMARY.

I have thus given a brief and imperfect sketch of the Agricultural Schools in Europe. Many facts, which would be important, it has been impossible for me to obtain in the limited time allotted me. I presume that the list which I have given, falls short of the actual number of such schools in Europe, for I doubt not that they have been increased a good deal within the last few months. Thus, it was announced some time since, as the intention of the Prussian government, to establish superior schools in all the provinces of that kingdom. And from a recent paragraph in the Cologne Gazette, I presume that this intention, delayed for a time from the disturbed political state of the country, has been carried into effect. That Gazette states, in its number for Sept. 27th, 1850, that "the government has ordered the establishment of agricultural schools, on scientific principles, on a very large scale."—*N. Y. Tribune of Oct. 24th*, 1850.

But though my list is doubtless deficient, I have been amazed, as I doubt not the committee will be, at its extent. The following summary will bring the whole subject under the eye :—

SCHOOLS,	Superior Schools.	Intermediary Schools.	Inferior Schools.	Special Schools.	Connected with Colleges and Universities.	Total.
In England, - - -	1	-	4	-	-	5
In Ireland, - - -	1	25	34	-	3	63
In Scotland, - - -	-	-	-	-	2	2
In France, - - -	5	-	70	-	-	75
In Italy, - - -	-	-	1	-	1	2(?)
In Belgium, - - -	-	3	2	1	3	9
In Prussia, - - -	3	2	12	13	2	32
In Austria, - - -	4	-	3	25	1	33
In Wurtemberg, - -	1	2	1	3	-	7
In Bavaria, - - -	1	1	32	1	-	35
In Saxony, - - -	1	3	-	1	-	5
In Brunswick, - - -	-	1	1	-	-	2
In Mechlenburg Schwerin, -	-	1	-	-	-	1
In Schleswig Holstein, -	-	2	2	-	-	4
In the Principality of Anhault, -	1	1	-	-	-	2
In the Grand Duchy of Hesse,	-	2	-	-	-	2
In the Grand Duchy of Weimar,	-	-	-	-	1	1
In the Duchy of Nassau, -	1	-	-	-	-	1
In the Electorate of Hesse, -	1	-	-	-	-	1
In the Grand Duchy of Baden, -	-	1	-	-	-	1
In the Duchy of Saxe Meiningen,	-	-	1	-	-	1
In Russia, - - -	2	10	51	4	1	68
Total, - - -	22	54	214	48	14	352

Though these schools have been grouped together in the above table into four distinct classes, yet in fact they are not always as distinct as thus represented ; and I have found it difficult, in all cases, to fix the place of a school on this scale. But these distinctions are in the main correct, and therefore of use, if understood as not mathematically accurate.

In the Tabular View of the Agricultural Schools of Europe, appended to this report, will be found many of the facts above detailed, and others that have not been named. The blanks in that table are, indeed, numerous; but the labor of preparing it, even in its present imperfect state, has not been small. Even a list of the places where these schools are situated, will, I trust, be of some importance, since, if made public, it might direct the attention of travellers to the schools, which otherwise they would pass unnoticed. And thus, since Americans are found yearly in so great numbers all over Europe, we may hope ere long to obtain full information concerning the schools. That travellers have hitherto, for the most part, been ignorant of their existence, is the only way in which I can explain the fact that so little is known of them.

State of Practical Agriculture in European Countries where Schools exist.

England and portions of Scotland doubtless exhibit the best examples of husbandry in Europe. In Ireland you see much in the cultivation to admire. Holland shows the great patience, as well as no small degree of success, of its farmers. Belgium has been too much celebrated, perhaps; yet its farmers have had much success upon tracts of land poor and sandy; but the principles of agriculture are not widely diffused, and the skill of the people is limited to a few points of husbandry. France certainly exhibits much excellent tillage to an unprejudiced eye, not generally, however, equal to that in England. Italy, with its fine climate and naturally good soil, cannot but show success, even under comparatively unenlightened tillage. In the valleys of Switzerland we see the influence of comparatively free institutions and natural ingenuity in the cultivation of the soil; so that the traveller is struck with the fine appearance of the farms, and their productiveness, even in some cases at the foot of the glaciers. There is a great difference, however, in the different cantons, and the difference corresponds in a good degree to the state of education. In Germany there is much difference in the agricultural condition of different states and kingdoms. In Prussia, the farmers have had to contend with a poor soil, and with feudal impositions. Yet one sees fine farming in the Rhenish provinces. Saxony, with a naturally rich soil, is said to exceed all other parts of Germany in her husbandry. In Bavaria and Wurtemberg, though abounding in agricultural societies and schools, the general state of agriculture is not elevated. In Russia, so thinly settled, there is a vast amount of uncultivated land. Yet in the vicinity of the cities, and where there are enough laborers, good cultivation is common. In this country is the famous *Tchornozem*, or black earth, a deposit, often from fifteen to twenty feet thick, so fertile as to need no manure, yet occupying an area " large as an European kingdom." (Murchisons's Geology of Russia, Vol. I, p. 557.) It seems to correspond essentially with the prairie soils of our country, and with the *regur*, or rich cotton soils of Hindostan. In Scandinavia the Swedes have shown commendable skill and industry in the cultivation of a country geologically analogous to that of New England. (*For further details, see Professor Johnstone's Syracuse Address.*)

Now it must be confessed that we do not find a very close correspondence between the practical agriculture of European countries, and the number of their agricultural societies and schools. For wherever schools exist, societies are also numerous. In France, for instance, there are nearly two hundred societies, and seventy-five schools. In Bavaria is a general society, numbering 8,000 members, and thirty-five schools. Hence, some would infer that schools are useless. But in the first place, if schools are useless, so are societies. Secondly, we must recollect that most of these schools have existed only a very few years, and that agricultural improvements are necessarily introduced slowly ; so that as yet we ought not to expect to witness much effect from these institutions. Thirdly, very much depends upon the manner in which the schools are conducted. If the community take little or no interest in them, no matter how able the professors. Now if, in this country, the government should take such an exclusive direction of agricultural schools as is done in France, Wurtemberg, and Bavaria, it would be almost certain that the people would take no interest in them. Fourthly, it cannot be doubted, that there are causes operating to advance or retard agriculture, more powerful than schools and societies. For example, natural fertility, or sterility ; wars and revolutions ; oppressive taxes ; individual ignorance, or knowledge. In England and Scotland, for instance, so elevated is the intelligence of the larger farmers and proprietors that they scarcely need schools, since each man is on the look out for improvements, and has the skill to apply them. I contend only, that schools are among the means of advancing the interests of agriculture, without attempting to fix their comparative importance among other means. For in such a country as ours, we need to use them all.

Remarks upon the Facts.

I. *They show us an extraordinary increase of interest in Europe, on the subject of Agricultural Schools, within a few years past.* Mr.

Colman finished his able Report on European Agriculture in 1844, and in it he describes only nine agricultural schools, though others then existed on the continent, of which he was not probably aware. But the larger part of the 352 above enumerated have come into existence since that time. The fact shows a strong conviction on the European mind, of their importance, and is one of the fruits of the peace that now generally prevails.

II. *These schools may be divided into two classes :—*

1. Professorships in colleges and universities.
2. Independent schools.

III. *The latter class, by far the most numerous, may be grouped into four kinds*, differing mainly as to the extent of instruction given, and the objects aimed at. 1. Superior ; 2. Intermediary ; 3. Inferior ; 4. Special. The last includes those devoted to instruction in some particular branch, such as horticulture, the cultivation and preparation of flax, irrigation, &c.

IV. *With a very few exceptions,—I do not recollect any, save the university of Edinburgh,—a farm, or at least a few acres of land, is connected with the school.* Thus the practical part of agriculture is attended to, in a manner, it would seem, that must satisfy those who have the least confidence in theory.

V. *The course of instruction, both scientific and practical, is essentially the same in all the schools of Europe.* Some include more branches than others, but there are certain essentials always included in the instruction.

VI. *European Agricultural Schools have taught us some important lessons.*

1. That these schools usually fail, if they do not receive efficient aid from the government.

2. That agricultural societies are not sufficient. These exist in all the countries above enumerated, I believe, and have done great good. But the conviction is very general there, that schools should be added to the societies; especially in countries where cultivation is a good deal deficient.

3. That theory is to be tested by practice ; and such theories as will not sustain this test are to be rejected.

4. That these schools are doing very much to promote the progress of agriculture. This was the general testimony.

5. That to teach agriculture in the primary schools and academies is not sufficient. This does some good, but does not accomplish all that is desirable.

6. That agricultural professorships, in colleges and universities, are

not sufficient. 1. Because lectures of this sort attract but few of the students of the colleges, who are looking forward to professional life. Such is certainly the case everywhere in Europe. 2. Because the two classes of students, who would thus be brought together, would have too little sympathy to act in concert, and as equals, in the same institution. 3. Because, without such concert and sympathy, one or other of the classes of students would feel no pride in the institution ; and without such an *esprit de corps* it could not prosper. 4. Because such professorships, unless numerous, would be entirely insufficient to accomplish the objects desired.

VII. *The lowest class of schools in Europe, the Inferior, are not needed in Massachusetts.*

1. Because the sons of our farmers are so well educated, that they can easily be prepared to enter at least the second grade of schools.

2. Because these inferior schools are adapted rather to paupers and orphans, and others in destitution, than to the sons of farmers in easy circumstances.

3. We already have institutions of this sort, as the Farm School, in Boston, for those who are destitute.

4. Because the subject of agriculture can easily be introduced into our common schools, so far as is necessary to learn the common principles of farming, or to fit boys to enter the higher institutions.

VIII. *We learn that those agricultural institutions succeed best which are started and sustained by the mutual efforts and contributions of individuals, or societies, and of the government.*

The schools in France, started by enterprising individuals, languished, and some of them failed, until the government lent an efficient hand. Very likely, the want of governmental patronage had something to do with the failure of the Hofwyl School, in Switzerland ; and it struck me that the Agricultural College of England, at Cirencester, languishes for the same reason. It is said that in Wurtemberg, the schools succeed well, because the government started and sustains them alone. I know too little of their circumstances and condition to throw light on the subject. But sure I am, that, in this country, the mutual exertions of the government and the people will be essential to success ; for this reason, if for no other, that here the people constitute the government : and if their representatives do not act in conformity to their wishes, their decisions will be reversed.

IX. *We learn, from European experience, that independent agricultural institutions are essential to accomplish the object which is aimed at.*

1. Because the field is wide enough to require such establishments.

The principles of agriculture are based upon a large part of the physical sciences; and it requires a good literary education to understand those sciences. No man can understand the *principles* of farming, who is not more or less acquainted with chemistry, anatomy, physiology, botany, mineralogy, geology, meteorology, and zoology; and then the practical part requires an extensive acquaintance with various branches of mathematics and natural philosophy. Many important principles of agriculture can, indeed, be taught in the primary schools, or academy; but there should be, somewhere, institutions of a higher character, entirely devoted to a thorough instruction and investigation of the science of the subject.

2. Because it demands extensive collections, of various kinds, in order to elucidate the principles of husbandry; enough, indeed, to belong to any scientific institution, and too many to form a mere subordinate branch of some institution with a different object in view.

3. Because the number of instructors must be so large, that they could not conveniently form an adjunct to some other institution.

4. Because the interests of agriculture are large enough to demand an institution definitely consecrated to their promotion. No other art is so important, and, I may add, no other is so difficult to be successfully cultivated; and, therefore, every means possible should be employed to render it assistance.

X. *Essentially the same reasons, and of greater force, exist for the establishment of Agricultural Schools in this country, as in Europe.*

1. It is the most ready and effectual mode of making farmers understand the principles on which good husbandry is founded.

Some have an idea that there are no such principles, and that the cultivation of the soil is a hap-hazard affair, and that guessing is as good a way as any, to secure good crops. I am aware that all the principles of managing land are not yet well settled, and that different circumstances often disappoint those who follow the best rules. But if there are no scientific principles on which husbandry is based, then one mode of farming is as good as another; a position which I am sure no reasonable man will take. And if one mode of tillage is better than another, there must be reasons for the difference, and those reasons are the very principles we are seeking after. Now it will be the leading object of agricultural schools to discover these principles, and to apply them in such a manner as to satisfy others that they are safe and valuable. Another object of such schools is, to detect and disprove, both theoretically and practically, any false principles and practices that may have been adopted by the community as true. The sons of farmers, who attend these schools, will learn to distinguish the true from the false principles, on this subject; and, on their return to the paternal roof, will communicate their knowledge, and apply it in practice.

2. These schools will furnish the most effectual means of introducing, among farmers, improvements in husbandry. Those who have the management of such schools will be on the watch for every valuable improvement, and be ready to make trial of any that are proposed. They will be in communication with similar schools, and with agricultural societies, in other lands, and thus learn, very early, whatever may aid the farmer, in regard to stock, implements of labor, new crops, and modes of culture. They will, therefore, form centres of information for the agricultural community, and, by going to them, the farmer may see, in their collections, new tools, grasses, grains, seeds, &c.

I know that agricultural societies and individuals can do a good deal to diffuse this information; but, as above remarked, they are not sufficient. At least, so think most of the countries of Europe, where such societies exist. The governments there feel the need of something more. England has, indeed, reached a very advanced state of agricultural improvement, with but little aid from schools. But it has taken a long time; and, at present, she is just the country that least needs the aid of schools, because her agriculture is in so advanced a state, and there are so many societies and educated individuals who are doing all they can to advance the interests of husbandry.

3. Agriculture, more than any other art, needs special help. First, because the principles on which it is founded are more difficult to be understood and applied than those of other arts. The processes of the farmer depend upon some of the most profound and subtle principles of chemistry, meteorology, physiology, and zoology; and so complicated are the causes of success or failure, that often the man who could solve a problem in fluxions, would be unable to unravel them. Secondly, because, for various reasons, other arts have always been more attractive, especially for youth, than agriculture; and hence, fewer efforts have been made to understand and apply its principles. In our country, at least, it is decidedly behind most of the other arts, although confessedly the most important.

4. A comparison of the state of agriculture in Europe with its condition in our country, shows the great need we have of agricultural schools, as well as societies and individual efforts. There are, indeed, several circumstances that lead an European to underrate the state of our agriculture. The great amount of loose stones, (called *drift* by the geologists,) that cover the surface in the Northern States, and

which our farmers have been able only to *begin* removing, gives the soil a most repulsive and ragged appearance to a European, save one perhaps from Scandinavia. Then, too, our uncouth wooden fences, which so often shelter a broad strip of bushes or weeds, in the place of beautifully trimmed hedges ; and our undrained and uncleared swamps ; the large amount of pasture ground, scarcely reclaimed, in many cases, from its native state ; and our entangled and untrimmed forests ;—all these and other circumstances turn off the mind of the foreigner from the real question at issue, viz., whether the acres of ground that are cultivated here, produce much less than the same amount in Europe. But, making due allowance for unfavorable biases from such causes, it must be merely an excess of national vanity for any one, who has travelled in Europe, to be very sensitive, if gentlemen from that quarter of the globe represent us as much behind the farmers on the other side of the Atlantic, in respect to the cultivation of our farms. I do not see how any one can pass over those countries, and not be struck with the difference. He will see many crops there, of great service, scarcely if at all known here. And several subjects, there well understood, are scarcely known or but little attended to here ; e. g., draining and irrigation. (See Colman's Report on European Agriculture, Vol. 2, p. 592.) Attention to their forests forms a part of their legislation, and of courses of lectures in their agricultural schools. But in Massachusetts, and still worse, in many other states, our beautiful native forests are often treated in a manner little better than vandalism ; and these ornaments to the landscape are fast disappearing, apparently with no consciousness that they are of any value, except the sum of money they will bring in the market, as wood or timber. Some of the county agricultural societies, in Massachusetts, have, however, I believe, turned their attention to this subject.

XI. *We have, in this country, several advantages over Europeans, for establishing Agricultural Schools.*

1. We have a more perfect state of civil and religious liberty. Even most of the freest of those nations have a state religion to support, and, therefore, have not religious liberty. I observed that this fact was a constant source of embarrassment in the schools ; and, in general, those schools in Europe are controlled so entirely by the government, as in this country would alienate the people from them.

2. The absence here of those broad distinctions of caste, which prevail nearly in every part of Europe. I am satisfied that this is one of the greatest obstacles to such schools in Europe. If a school is open for the poor, then the rich, and even those but little above poverty, will stay away. If it is adapted to the wealthy, the poor cannot enter.

Some of the schools have changed their policy several times, in respect to this point. But among us there will be no difficulty in adapting a school to all classes. Indeed, if not so adapted, it would not be patronized; or, at least, would do very little good.

3. We are comparatively free from taxes, and can therefore be more liberal towards schools. Till he visits Europe, an American has little idea of the vast difference between men, in this respect, on the opposite sides of the Atlantic, nor on what vantage ground it places his countrymen; and how much easier it is for them, especially the poorer classes, to be liberal towards worthy objects, than in Europe.

4. In this country, the cultivators of the soil are usually the owners of it; and it is they, mainly, who must establish an agricultural school, if one is established, for their own and their children's benefit. But in Europe, for the most part, the laborers have little concern in the matter. They rarely own the soil which they cultivate; and, therefore, take but little interest in getting up schools: nay, they are sometimes afraid of them, lest they should so improve agriculture as to diminish their wages. The schools, therefore, are got up by those who own extensive estates, and who are anxious to make them more profitable; or by political men, who know how much the prosperity of a kingdom depends upon its agriculture.

5. The superior state of education in Massachusetts, among the laboring classes, gives us another decided advantage over Europeans in the establishment of agricultural schools.

It would be quite practicable, I think, to require, for admission to our schools, as much knowledge as is demanded in many of the European schools for completing the course; and that, too, if our pupils were the sons of the common farmers. Now, it is not national vanity, but simply the statement of a fact, to say, that this could not be done in any European country, save perhaps in Scotland, where few agricultural schools exist. In most other countries, the sons of the common peasantry, or laborers, are incapable of receiving that benefit from these schools, which the poorest might derive in this State. Here, they could start from a much higher point in their career of agricultural learning, and, of course, terminate their course at a level proportionably more elevated.

6. We have European experience to direct us in this matter, and can more easily avoid the mistakes which that experience points out, and introduce its better plans, than can be done by institutions already established.

XII. *The conclusion to which my own mind has been brought, by these investigations, is, that Massachusetts should lose no time in taking hold of the business of agricultural education, efficiently and liberally.*

I can hardly doubt, from all that I have seen and heard for the last ten years, that there is a strong desire and expectation that the Legislature should move in this matter without delay. I have heard it talked about almost everywhere; it has formed the burthen of a large part of the addresses delivered at the annual agricultural fairs; individuals have attempted to establish private schools; and at least two gentlemen, at their decease, have left splendid legacies for founding such schools,—one at Roxbury, and the other at Northampton. I refer to Hon. Benjamin Bussey, and Oliver Smith, Esq., of Hatfield. How soon the legacy of the former can be brought into use, for a school at Roxbury, I know not; but that of Mr. Smith will not be available for more than half a century. Yet let no one fear that we shall have too many schools of this description. Those countries in Europe where they are most numerous, are most inclined to multiply them; and my own conviction is, that, ere fifty years are gone by, the State will find that she needs several of these institutions, of the superior class. I apprehend that it will not be found wise to allow the number of pupils in any of them to exceed two hundred.

Perhaps here I should rest; but so naturally have I been led to inquire into the proper character of agricultural education in this State, should any be adopted, that I shall venture to obtrude upon the committee my conclusions. I trust, however, that I shall not be so tenacious of these views, as not to be willing to adopt better ones, which other members of the committee may suggest.

The first question that arises, is, *What are the objects we may hope to attain by founding agricultural schools in the State?*

I reply, and here I must be pardoned for a repetition of some sentiments,—

1. First, such schools would furnish to all classes of the community, an opportunity to acquire a definite knowledge of all the known principles by which agricultural pursuits should be conducted. These principles are the joint results of all the experience men have ever had in husbandry, and of all the deductions they have made from the sciences on this subject. That there is now a great deficiency of such knowledge in our community, no man in his senses will deny. Many of the principles of agriculture are, indeed, themselves yet unsettled; but a great many are settled, of which our farmers are yet ignorant, or know them only imperfectly. To explain and illustrate these, will be the grand business of agricultural schools. Hence, it is the sons of farmers who would be most interested in attending them. But there are many others, who desire to prepare their sons for agricultural pursuits; at least, to enable them to oversee farming establishments, if not

to labor with their own hands; and they would find such schools a most important auxiliary to their success.

2. In such schools our youth would find an excellent opportunity of learning the best method of conducting the practical operations of the farm; in other words, of seeing theory reduced to practice. They would learn, also, how to labor with their own hands, for I take it such a requisition would be indispensable in these schools. And thus might we hope that practical farming would become more popular, as it should be, among all classes of the community.

3. These schools would form centres of information on the subject of agriculture, and from them our farmers could derive important aid. They would keep in communication with similar institutions throughout the world, and thus would learn all that might be new or important in husbandry, and spread it through the community.

4. These schools would be the best places for testing the value of supposed improvements in agriculture. Such improvements, coming accidentally to the farmer's ears, are now often tried, and not unfrequently found of no value, and this is one of the circumstances that prejudices farmers against what they call book farming. But let such supposed discoveries be first tried on the experimental farm connected with the school, and their deficiency or value may be thus learnt once for all.

5. They ought to be places for making improvements in agriculture. We ought not to expect too much in this respect, especially when the schools are new. But as the instructors become experienced, and the means in their hands more ample, it may be hoped that they will push their researches and experiments into unexplored regions, and bring back some fruit for the practical farmer.

6. Finally, to sum up the whole in one word, the grand object of these schools is to improve the husbandry of Massachusetts. It is to make the acres that now produce something, produce more, and thus to stimulate the rising generation to reclaim more acres from the waste condition in which they now lie. One is struck, in travelling through the state, to see how large a part of its surface produces next to nothing, although capable of yielding a rich supply; nay, in how many cases the richest portions of its soil,—its swamps, for instance,—remain in their native condition. Nor is it strange that our sons, when they see so little to encourage them at home, should brave the hardships and the fevers of the western prairies, in order to enjoy their prolific soil. Agricultural schools will aid in checking this disposition to emigration, and tend to multiply the cultivated acres of Massachusetts. Along with agricultural societies and individual efforts, which will still be needed

as much as ever, we may hope these schools, properly founded, and judiciously managed, will serve to bring agriculture more nearly upon the ground occupied by the other arts. Now it is decidedly below them; and yet these arts can never thrive long if agriculture is neglected, and, therefore, it is for the interest of all classes to join in efforts to advance the cause of agriculture.

But what shall be the character of the Agricultural Schools, should any be established in Massachusetts? This difficult inquiry I would meet by a few suggestions. But after the survey that has now been taken of agricultural education in Europe, it cannot be expected that I should recommend, or that the community will be satisfied, with schools of an inferior class. Twenty, or even ten years ago, it might have answered to propose the introduction of agriculture into our primary schools, or as a department in our academies, or a professorship in our colleges. All this it may be well enough to do now, but something more must be done. So Europeans judge, and accordingly, as we have seen, they have started institutions with as ample a foundation, and as numerous a body of instructors, as we find in most of our American colleges. Nothing short of this, as it seems to me, will be sufficient for our country; nay, I fancy that at least one such superior institution is needed in each of our states. The work to be done is too great, the number of teachers is too many, and the amount of various collections too large, to attempt to attach an agricultural school to some other institution, and that too as only a subordinate branch. Even if agriculture is taught in our colleges, academies, and primary schools, it needs some one institution devoted entirely to the subject, to give effect and completeness to the subordinate teaching, and to carry it still farther; otherwise the agricultural knowledge will be as the literary would be, if the universities and colleges of the land were taken out of the way, and only the primary schools and academies remained.

But though our community, as I believe, especially the intelligent farmers, are prepared to appreciate the importance of such superior institutions, I fear that but few are ready to devote the amount of funds requisite for putting such a seminary at once into full operation. Nay, none but those who have had experience know how large an amount of money it requires, with the strictest economy, to found and carry on successfully a large institution of learning. My hope is, however, that the government and the people will start this enterprise, if they do it at all, with a high standard in view, even though they cannot, till a considerable period, reach the height of their wishes. It seems to me that to aim much lower than the plan which follows, will be likely to make the whole subject, ere long, contemptible, or at least a failure.

Outlines of a Plan for Agricultural Education in Massachusetts.

I. Let an Agricultural School or College of the superior class be established somewhere in the State, possessed of all the means (teachers, books, apparatus, specimens, farms, &c.) necessary to give a finished education in the principles and practice of agriculture. Such a school, it seems to me, should embrace the following particulars, at least :—

1. *A school of instruction, by lectures and recitations, in the following branches :*

a. Practical agriculture.

b. Chemistry, elementary and applied.

c. Natural history, especially zoology and botany.

d. Elementary and applied mineralogy and geology.

e. Anatomy and physiology, human and comparative.

f. Veterinary medicine and surgery.

2. *Collections of the following description :—*

a. Models of agricultural instruments.

b. Of dried seeds.

c. Of dried grasses, grains, &c., entire.

d. Specimens planed, of the useful kinds of wood.

e. A pomological collection, or models of the varieties of fruit.

f. Small collections of simple minerals and rocks for elementary instruction in mineralogy and geology.

g. An economic collection, embracing all the minerals, rocks, fossils, soils, marls, clays, &c., useful in the arts, exhibiting them in their various stages of preparation, with specimens of the finished articles, constituting what in Europe is called a museum of economic geology.

I saw three such collections in Europe ; one at the School of Mines in Paris ; one in London, the fruit of the Ordnance Survey ; and one in Dublin, having the same origin ; the two latter not yet opened to the public, but very choice and splendid collections. They must be very serviceable for all the arts.

h. Insects injurious to vegetation.

i. Stuffed specimens and drawings of the species and varieties of animals useful in agriculture.

k. A museum of human and comparative anatomy, including a manikin.

l. Chemical and philosophical apparatus.

m. A scientific and agricultural library.

Some of these collections are more important than others; but they would all be of service, and tend to give permanence to the institution. To procure and arrange them would require a long time, but they might be begun at once. I saw all of them, larger or smaller, in some of the agricultural schools of Europe, though nowhere at the same school. Indeed, many of those schools appeared to me to be quite deficient in collections, and I thought this one cause why they were in a decaying state. When literary institutions are called, as they often are, in the early part of their course, to pass through seasons of storm, good libraries, and ample apparatus, and collections, form one of the best of anchors to enable the vessel to outride the tempest.

3. *A model and experimental farm of moderate size, and instruction in practical farming.* I see no use in a large farm, as in general a small one, say of 100 or 200 acres, will embrace every important variety of soil, and can be more easily managed. It seemed to me that some of the European schools have farms so large as to be unwieldly.

4. *Provision for Instruction in Ancient and Modern languages.* Such studies should not be required. But some might wish to get a little knowledge of Latin and Greek, in order to become better naturalists, and understand better the scientific terms in agriculture; and others would wish to acquire French or German. Such instruction would be no cost to the State, as teachers might be found near at hand, most probably, in almost any part of the State, and it would render the school more attractive and respectable. It has been seen that such instruction is provided in some of the schools in Europe. It might be well, also, to provide similar instruction in the higher mathematics, as is done frequently in Europe.

5. *Provision on the farm for the board of those students who are willing to devote their time to labor daily beyond what is required of all.* For these extra labors, in my opinion, a compensation should be given, such as should at least meet the charge for board. This might enable many poor but worthy young men to enjoy the benefit of the institution, who would otherwise be deprived of the privilege.

6. *Number of Instructors Necessary.*

1. A professor of horticulture, sylviculture, and rural legislation, who should also be chairman of the board of instructors, or president of the institution.

2. A professor of agriculture.

3. A professor of elementary and agricultural chemistry.

4. One of natural history and geology, who should be curator of the collections.

5. One of anatomy, physiology, and veterinary medicine and surgery.

6. One of the mathematics of agriculture, such as farm accounts, irrigation, draining, surveying, levelling, construction of roads, bridges, &c.

This appears to me the smallest number of professors with which an institution could be respectable and useful, even at its commencement. The number is much less than it is at nearly all the higher agricultural seminaries in Europe. There it ranges from eight to twenty.

Besides the professors, there should be a superintendent of the farm, and of all practical operations of the establishment, who might also give some lectures, say on practical agriculture.

If one of the professors should be a clergyman, he might act as chaplain of the establishment.

7. *The course of study should embrace two years.* When creditably finished, the pupil should be entitled to an agricultural diploma or degree.

8. *For admission, an examination should be required, in English grammar, geography, arithmetic, and Euclid's Elements of Geometry,* at least the first five books. This may seem a high standard. It is higher than most of the schools in Europe. But the young men of Massachusetts, even the poorest, can easily come up to it, and thus make their subsequent course much more profitable. One great complaint in Europe is, that the pupils are unable, for want of early education, to understand the instruction.

9. *The tuition should be as low as possible, certainly not higher than at the existing colleges of the State,* and that is about $40. As soon as possible the instruction should be gratuitous.

10. *Several scholarships, say as many as ten to begin with, equal at least to the tuition, should be founded by the State,* to be given to the most diligent and successful pupils. Also, several prizes of a smaller amount should be offered.

This is the almost universal practice in Europe, and its operation is excellent.

11. *Provided individuals shall offer ten acres of good land, to be improved as a model farm, by some academy, let the State give to that academy* (not more than to one in each county,) *at least* $200 *for instruction, and* $50 *for a library; provided said academy shall agree to cultivate the land, and provide for a department of agriculture: the aid from the State to be withdrawn, however, when the number of students in agriculture shall be less than ten.*

12. *Let a manual of agriculture be prepared by some competent person, or some existing manual be adopted, and introduced into the primary schools, if any children wish to study it.* It might be well for the State to furnish the books gratis.

13. *One object of these schools of a lower grade should be to prepare pupils to enter the central institution, in advanced standing, if fitted for it.*

14. *By the addition of a single professorship of technology to such an institution as has been described, and extending the collection of instruments to those of every art, this school might become a school of mines, as well as of commerce and manufactures, and thus afford an education to the son of the mechanic and merchant, as well as the farmer.*

I do not, however, recommend that such an addition should be made at the outset, lest by aiming at too many objects, the whole be lost, as we have seen to have been the case sometimes in Europe.

15. *Let the State appoint a Board and Secretary of Agriculture, who shall sustain the same relations to that interest and the schools connected with it, as the Board and Secretary of Education do to primary schools.*

Methods by which the State can put the preceding Plan into Operation.

1. The first method would be for the State to assume at once the whole expense and responsibility of the entire system. For the first year the expenses would be nearly as follows :

Purchase of a farm of 200 acres, with farm house, barns, &c., say,	$10,000
Boarding house, (if necessary,)	10,000
Building for lectures, recitations, and cabinet,	25,000
Apparatus, books, models, specimens, &c.	10,000
To establish agricultural departments in the academies in different counties, say,	3,000
For scholarships, prizes, and other contingent expenses, say,	5,000
Salary of a president,	1,200
Do. of five professors,	5,000
Do. of superintendent of the farm,	800
	$70,000

If we suppose there would be one hundred students, who are charged $40 each for instruction, it would furnish $4000 towards the expense, reducing it to $66,000. I have not supposed that the farm and boarding establishment would do more than to support those who manage them for the first year or two. Perhaps little or nothing should be expected from them subsequently, more than this.

If the State should choose to endow at once the presidency, and five professorships, it would require an additional appropriation of about $105,000. But I have supposed the State would prefer endowing them prospectively, whenever funds should accrue, from the sale of lands in Maine or other sources, that would not be needed for other purposes.

On this supposition, it would be necessary for the State to pay the salaries of the officers of the school, until such funds should become available. After the first year, then, we might calculate the annual expenses as follows :

For salaries of president and professors, - - -	$6,200
Do. of superintendent of the farm, - - -	800
For scholarships, prizes, collections, and contingencies, -	3,000
For aid to county schools in the academies, - - -	3,000
	$13,200
Deduct tuition of 100 pupils, - - - - -	4,000
Annual expense after the first year, - - - -	$9,200

If, now, the government of the State were prepared to vest the above sums in this enterprise, they might place the institution on a solid basis, and bring it into operation without delay. I think a less amount could not accomplish the objects in view. It is far below the sum recommended for the same purpose, by an intelligent committee of the legislature in a neighboring state. It is not greater than the farmers of the State, if I rightly judge of their feelings, would approve; certainly not greater than their interests deserve. Yet not improbably many will feel that it would not be wise for the legislature to make so large an appropriation at the beginning of the enterprise, lest it should fail, or the public mind is not prepared for it. I beg leave, therefore, to suggest another method, by which a considerable part of the expenditure might for a time be saved, the school brought even earlier into operation, and ultimately reach the same standard, as in the first plan. This method is as follows :

2. *Let the agricultural institute be located so near some existing literary institution, that the pupils could attend its scientific lectures and study its collections in natural history.*

By such a plan, the pupils of the agricultural school might have access almost, perhaps quite, free of expense (unless so numerous as to render it necessary to enlarge the lecture rooms,) to the following courses of lectures, until the State should be ready to endow a full corps of professors, and make the proper collections, when the temporary connection between the two institutions could be dissolved:

1. A course on anatomy and physiology.
2. On mineralogy.
3. On geology.
4. On zoology.
5. On elementary chemistry.
6. On botany.
7. On natural philosophy.

Such a course as this, if desirable, is entirely practicable, as I am prepared to show, if the committee wish for details.

On this plan, the expenses would be as follows:

For a farm and boarding house, as before, - -	$20,000
For lecture rooms and cabinet, - - - - -	25,000
For apparatus, books, and specimens, - - -	10,000
For agricultural departments in academies, - -	3,000
For scholarships, prizes, and contingencies, - -	5,000
Salary of a president, and professor of agriculture, -	2,200
Do. of a superintendent of the farm, - - - -	800
For extra courses of lectures by gentlemen from abroad, say on agricultural chemistry, and veterinary medicine, and surveying, say, - - - - -	800
	$66,800
Tuition of 100 pupils, - - - - - -	4,000
	$62,800

On this plan, only two professors need be directly and exclusively connected with the school, and the annual necessary expenses, after the first year, would be as follows, until the State should choose to separate it from the literary institution:

Salary of president and professor of agriculture, - -	$2,200
Do. of superintendent of the farm, - - - -	800
For the academies, - - - - - -	3,000
For scholarships, prizes, &c., - - - - -	1,000
For extra lectures, - - - - -	800
	$7,800
Tuition of 100 pupils, deduct, - - - -	4,000
	$3,800

As to the immediate or prospective endowment of the presidency and a professorship of agriculture, the State could do as it judged expedient, as in the first plan.

It will be seen, that, by this second plan, the principal reduction of expense would be in the annual outlay after the first year. Yet the farm and buildings upon it would be the property of the State, although their sale for other purposes might involve some loss. But after all, there will be some, I doubt not, perhaps many, who will feel that in a new enterprise like this, and untried in our country, the State would not be justified in expending upon it a sum so large as even this second plan requires. They are in doubt whether the great body of the people are prepared for it, and if the government were to take the lead in this matter and get up a large establishment, it might be that the people would not sustain it by sending their sons to it. They would prefer, therefore, that individuals should take the lead in the enterprise, or, at any rate, that the government should only offer certain sums for the object, provided individuals will furnish equal sums. Thus public feeling would be tested, and if the school should be established, a deeper interest would be felt in it, than if founded by the government alone.

There is certainly much truth and wisdom in such suggestions, and in view of them, I would present a third method, in which I think a school might be established, which would meet essentially these views, and yet not give up the hope of ultimately reaching the high standard proposed in the two first plans.

Third Plan for an Agricultural Institute.

I. *Let a Board of Agriculture, as in the first plan, be appointed by the Legislature, with the following, among other powers* :

1. To appoint a secretary, one of whose duties for the present should be, to aid in the establishment of agricultural schools, and he should also be expected to give a course of lectures in the central school, should such a one be established and his services be required.

2. To appoint a president of such central school or college, with a salary of **$1,200**, who should be also a professor of some department in the same ; also, to appoint a professor of agriculture in the same, with a salary of **$1,000**. Also, a superintendent of the farm, with a salary from **$800** to **$1000**.

3. These four officers, the president, secretary, professor of agriculture, and superintendent, should have power, with the board, to select a site for such college, and purchase a farm, containing from 100 to 200 acres, provided the following conditions could be realized.

1. If the school can be located near enough to some existing literary or scientific institution, for the pupils to attend its lectures, and study its collections of natural history, it being understood that admittance to those lectures and cabinets shall be gratuitous, or nearly so, and that the lectures embrace the following subjects : 1. Chemistry ; 2. Botany ; 3. Mineralogy ; 4. Geology ; 5. Zoology ; 6. Anatomy, and physiology ; 7. Natural philosophy.

2. If the friends of agricultural education in the place where the school is to be located, or in other places, shall raise a sum for the purchase of a farm, and sustaining the school, equal to the sum appropriated by the State, which sum shall not exceed **$20,000**, nor be less than **$15,000**; that is, the State shall not be called on to pay over **$20,000** in any case to establish and carry the school through the first year.

4. *The Board should have power to procure for the school, annual courses of lectures, upon subjects not provided for by the instruction of the president, secretary, professor of agriculture, and superintendent, nor by the lectures in the adjacent literary institution ;* as for instance, the subject of veterinary medicine and surgery, and agricultural chemistry.

5. *Power to appoint one tutor for every thirty-five scholars in the school, with a salary of* **$500**, whose duty it shall be to hear recitations in all the branches studied in the school, as well as to examine the classes upon all the lectures they hear, and, in connection with the president, professor of agriculture, and superintendent, to watch over the conduct of the scholars, and exercise all the discipline of the school.

6. *Power to make an offer to the public, in behalf of the State, of the prospective endowment of a professorship of elementary and agricultural chemistry ; of natural history and geology ; of anatomy, physiology, and veterinary medicine and surgery,* whenever funds from the sale of lands in Maine, or from other sources, not demanded for other objects, shall come into the possession of the State ; provided that individuals or societies shall agree to give one half of said endowments, with the privilege of attaching their names to the professorships ; and provided also, that a professorship shall not be reckoned higher than **$20,000.**

7. *Power to expend* **$3,000**, *the first year, in establishing agricultural departments in as many of the existing academies of the State as shall be judged expedient ; provided,* that from eight to ten acres of good land shall be provided, to be cultivated as a model farm by said academy ; and provided the number of scholars in the academy, studying agriculture, shall not be less than ten.

8. *Power to make such other arrangements for the management of the*

new school, as they may judge necessary ; particularly to allow such students as cannot be accommodated with board at the farm house, to take rooms in the town where the school is situated, and, in renting the farm, see to it that opportunity be given to those pupils, who may desire it, to pay for their board by labor.

9. *Power to invite donations from the citizens of the State, of models and drawings of agricultural instruments, and drawings of animals useful in agriculture ; models of varieties of fruit ; specimens of insects injurious to vegetation ; of the skeletons of animals ; of dried seeds, grasses, and grains ; of rocks, minerals, and soils ; of agricultural and scientific books, and whatever substances, organic or inorganic, may be useful in such a school.*

10. *Power to invite the friends of agriculture to establish scholarships and prizes in the central school and in the county schools.*

11. *Power to secure the preparation and introduction into the primary schools of a proper manual of agriculture, to be taught there to those children whose parents wish them to attend to that subject.*

12. *Were such an arrangement to go into operation, in which the government defrayed half the expense, and individuals the other half, then should the central and the county schools be under the control of trustees, one half of whom should consist of the board of agriculture, and the other half of gentlemen chosen by the donors to the several institutions ;* one hundred dollars to constitute a share, and to entitle the donor of that sum to one vote for trustees, two hundred to two votes, and so on.

Expenses of the School upon the above Plan for the first year.

I will suppose the donations of individuals for founding the school in the place where it is located, and in other places, to rise no higher than $15,000, although we might reasonably hope, I think, to raise it to $20,000. Adding to this the appropriation of the State, we should have $30,000. If we suppose only 100 students for the first year, to pay $40 each, we must add $4,000 to $30,000, equal to $34,000, the sum with which to found the school. Let us see whether this sum will be at all adequate to put the school into operation.

Cost of a farm from 150 to 200 acres, with farm houses, and a boarding house,	$20,000
For agricultural departments in the academies, if equal sums are raised by individuals,	3,000
Salary of a president,	1,200
Do. of a professor of agriculture,	1,000

Salary of a superintendent of the farm, who should also give lectures, - - - - - - -	1,000
Salaries of three tutors, (one for every thirty students,) -	1,500
Extra courses of lectures, - - - - - -	600
Amount,	$28,300
Funds on hand, - - - - - -	$34,000
Excess of funds, - - - - - -	5,700

This excess would doubtless be needed for contingencies that must be met, especially the first year. Some of it, certainly, would be wanted for a library. I have not supposed any income from the farm. Probably there might be none the first year, as it would most likely require two head farmers, under the superintendent, to carry it on, and doubtless many improvements would need to be set on foot.

Cost of the School in subsequent years.

For the academies, - - - - - -		$3,000
Salaries of president, a professor, and a superintendent, -		3,200
Salaries of three tutors, - - - - -		1,500
Extra courses of lectures, - - - - -		600
		$8,300
150 students at $40 each, - - - -	$6,000	
Income of the farm, - - - - - -	200	
		$6,200
Annual expense after the first year, - - - -		$2,100

So far as the number of students and the income of the farm are concerned, the above statements are hypothetical; that is, the estimates are based only on probabilities, and may be quite erroneous, although I have endeavored to place the numbers very low. But as to the salaries, they are put even higher than they now are at the literary colleges in the country, and I feel confident that competent men can be obtained for those sums. I am quite sure, also, that for $20,000, a farm of 200 acres, with farm houses, and a boarding-house, can be purchased in a situation which would answer the other conditions of this plan. I mention these facts to show that the plan is a practical one, although it would be premature for me to go into details, which very probably may never be wanted. If they are needed, I am ready to furnish them.

In conclusion, I would only say, that my conviction is, that if Massachusetts desire it, she might, on some such plan as has been suggested, within a very few months, put into operation a system of Agricultural Schools that would ultimately attain a very high character. Some will doubtless think the standard to be too high. But it is not higher than is common in Europe. If so many examples of liberality were not before us in that quarter of the globe, or if inferior schools alone had not there been found insufficient, we might be justified in adopting some narrow system of instruction. But I fear that such a course would disappoint its friends, and confirm the prejudices of its enemies. I cannot believe that the citizens of Massachusetts, especially its farmers, will be content with a narrow system. Their common schools are too well known all over the world to allow them to fall far below Europeans in their Agricultural Schools. Is it not high time to show that the State, which has distinguished herself so much by the literary education of her children, can rise equally high in her agricultural system of education?

We hear men speak of the great expense of agricultural schools, as if the money bestowed upon them were lost, or, at least, unproductive. But if, as we have reason to believe, those schools would sensibly advance the cause of farming in the State, our money could not be better vested. For to advance our agriculture is to give a new impulse to all other industrial pursuits; nay, to all our great interests, economical, political, social and moral. Money expended for such an object, is, therefore, put out at compound interest; and our children, if not ourselves, will reap the rich reversion.

All which is respectfully submitted,

EDWARD HITCHCOCK.

AMHERST, *Dec. 16th*, 1850.

A TABULAR VIEW

OF THE

AGRICULTURAL SCHOOLS OF EUROPE

IN 1850.

A TABULAR VIEW OF THE AGRICULTURAL SCHOOLS OF EUROPE IN 1850.

ENGLAND.

Locality of the School.	Year when established.	Rank of the School.	Character of the Instruction.	Price of Tuition and Board.	Age of Pupils.	Length of the Course.	Number of Professors.	Number of the Courses.	No. of Students. Now in school.	No. of Students. From begin'g.	Paid by Government.	Scholarships.	Acres in the Farm.
Cirencester, - - -	-	Superior.	College.	353	14 to 20	3 yrs.	6	7	50	-	-	-	700
Hoddesdon, Hertz, - -	1845	Inferior.	Training School.	233	-	-	6	5	50	-	-	-	300
Kimbolton, - - -	-	do	Agricultural Grammar School.	111 to 133	-	-	-	-	-	-	-	-	-
Kennington, - -	-	do	Scientific Agricultural Academy.	-	-	-	-	-	-	-	-	-	-
York, - - -	-	do	Training School.	-	-	-	-	-	-	-	-	-	-

IRELAND.

Locality of the School.	Year when established.	Rank of the School.	Character of the Instruction.	Price of Tuition and Board.	Age of Pupils.	Length of the Course.	Number of Professors.	Number of the Courses.	No. of Students. Now in school.	No. of Students. From begin'g.	Paid by Government.	Scholarships.	Acres in the Farm.
Glasnevin, - - -	1845	National Model Agr. School.	Theoretical and Practical Agriculture.	-	-	-	-	-	43	-	-	-	128
Larne, Antrim Co. - -	do	Model School.	do	55	-	2 to 3	-	-	-	-	-	-	Not over 8
Markethill, Armagh Co. -	do	do	do	-	-	-	-	-	-	-	-	-	do
Hollywood, Doun Co. -	do	do	do	-	-	-	-	-	-	-	-	-	do
Carrick, Fermanagh Co. -	do	do	do	-	-	-	-	-	-	-	-	-	do
Longhash, Tyrone Co. -	do	do	do	-	-	-	-	-	-	-	-	-	do
Sallybank, Clare Co. -	do	do	do	-	-	-	-	-	-	-	-	-	do
Belvoir, Clare Co. - -	do	do	do	-	-	-	-	-	-	-	-	-	do
Rahan, King's Co. - -	do	do	do	-	-	-	-	-	-	-	-	-	do

Loughrea, Galway Co. -	1845	Model School.	Theoret. and Prac. Agr.	-	-	-	-	-	-	-	-	-	Not over 8
Ballynakill, Galway Co. -	do	do	do	-	-	-	-	-	-	-	-	-	do
Kyle Park, Tipperary Co. -	do	do	do	-	-	-	-	-	-	-	-	-	do
Bailieborough, Cavan Co. -	do	do	do	-	-	-	-	-	-	-	-	-	do
Dunmanway, Cork Co. -	do	do	do	-	-	-	-	-	-	-	-	-	do
Dunlewy, Donnegan Co. -	do	do	do	-	-	-	-	-	-	-	-	-	do
Bath, Monaghan Co. -	do	do	do	-	-	-	-	-	-	-	-	-	do
Mount Trenchard, Lim'k Co.	do	do	do	-	-	-	-	-	-	-	-	-	do
Tervoe, Limerick Co. -	do	do	do	-	-	-	-	-	-	-	-	-	do
Ardfinnan, Tipperary Co. -	do	do	do	-	-	-	-	-	-	-	-	-	do
Derrycastle, Tipperary Co. -	do	do	do	-	-	-	-	-	-	-	-	-	do
Woodstock, Kilkenny Co. -	do	do	do	-	-	-	-	-	-	-	-	-	do
Leitrim, Leitrim Co. - -	do	do	do	-	-	-	-	-	-	-	-	-	do
Glandore, Cork Co. - -	do	do	do	-	-	-	-	-	-	-	-	-	do
Farraghy, Cork Co. - -	do	do	do	-	-	-	-	-	-	-	-	-	do
Belfast, - - -	do	do	Professorship in Belfast College.	-	-	-	-	-	-	-	-	-	
Cork, - - -	do	do	do	33	-	2	5	-	-	-	-	4 of 67	180
Galway, - - -	do	do	do (?)	-	-	-	-	-	-	-	-	-	
Ballinashee, Antrim Co. -	do	Ordinary Agricultural School.	Practical Agriculture.	-	-	-	-	-	-	-	-	-	2 to 3
Ballymena, Antrim Co. -	do	do	do	-	-	-	-	-	-	-	-	-	do
Dundrod, Antrim Co. -	do	do	do	-	-	-	-	-	-	-	-	-	do
Ballinahone, Armagh Co. -	do	do	do	-	-	-	-	-	-	-	-	-	do
Drumkerl, Cavan Co. -	do	do	do	-	-	-	-	-	-	-	-	-	do
Cloghan, Donegal Co. -	do	do	do	-	-	-	-	-	-	-	-	-	do
Ballaighan, Donegal Co. -	do	do	do	-	-	-	-	-	-	-	-	-	do
Ballyrashane, L. Derry Co. -	do	do	do	-	-	-	-	-	-	-	-	-	do
Ballyoughry, L. Derry Co. -	do	do	do	-	-	-	-	-	-	-	-	-	do
Five Mile Town, Tyrone Co.	do	do	do	-	-	-	-	-	-	-	-	-	do
Drumnasern, Tyrone Co. -	do	do	do	-	-	-	-	-	-	-	-	-	do
Dressog, Tyrone Co. -	do	do	do	-	-	-	-	-	-	-	-	-	do
Kilkishin, Clare Co. -	do	do	do	-	-	-	-	-	-	-	-	-	do

IRELAND—CONTINUED.

Locality of the School.	Year when established.	Rank of the School.	Character of the Instruction.	Price of Tuition and Board.	Age of Pupils.	Length of the Course.	Number of Professors.	Number of the Courses.	No. of Students.		Paid by Government.	Scholarships.	Acres in the Farm.
									Now in school.	From begin'g.			
Parteen, Clare Co. - -	1845	Ordin'y Ag. Sch.	Practical Agriculture.	-	-	-	-	-	-	-	-	-	2 to 3
Knocknaguihy, Clare Co. -	do	do	do	-	-	-	-	-	-	-	-	-	do
Tiernaboul, Kerry Co. -	do	do	do	-	-	-	-	-	-	-	-	-	do
Killoughteen, Limerick Co. -	do	do	do	-	-	-	-	-	-	-	-	-	do
Two-Mile House, Kildare Co.	do	do	do	-	-	-	-	-	-	-	-	-	do
Kilberry, Kildare Co. -	do	do	do	-	-	-	-	-	-	-	-	-	do
Ooning, Kilkenny Co. -	do	do	do	-	-	-	-	-	-	-	-	-	do
Ballyglass, Kilkenny Co. -	do	do	do	-	-	-	-	-	-	-	-	-	do
Dromisken, Louth Co. -	do	do	do	-	-	-	-	-	-	-	-	-	do
Ratoath, Meath Co. - -	do	do	do	-	-	-	-	-	-	-	-	-	do
Girley, Meath Co. - -	do	do	do	-	-	-	-	-	-	-	-	-	do
Carnaross, Meath Co. -	do	do	do	-	-	-	-	-	-	-	-	-	do
Clonmellon, Clonmellon Co. -	do	do	do	-	-	-	-	-	-	-	-	-	do
Ballinvally, Westmeath Co. -	do	do	do	-	-	-	-	-	-	-	-	-	do
Delgany, Wicklow Co. -	do	do	do	-	-	-	-	-	-	-	-	-	do
Esker, Galway Co. - -	do	do	do	-	-	-	-	-	-	-	-	-	do
Castlehacket, Galway Co. -	do	do	do	-	-	-	-	-	-	-	-	-	do
Galway Workhouse, Gal. Co.	do	do	do	-	-	-	-	-	-	-	-	-	do
Carawallen, Leitrim Co. -	do	do	do	-	-	-	-	-	-	-	-	-	do
Geevagh, Sligo Co. - -	do	do	do	-	-	-	-	-	-	-	-	-	do
Upper Arigna, Sligo Co. -	do	do	do	-	-	-	-	-	-	-	-	-	do
Templemoyle, - -	1827	Intermediary.	Theoretical and Practical.	-	-	-	-	-	70	497 to yr. 1844	-	-	169
Brookfield, - - -	do	do	do	-	-	-	-	-	-	-	-	-	-

SCOTLAND.

Edinburgh, - - -	1845	-	Professorship in Univer'y.	-	-	-	-	-	-	-	-	-	-
Aberdeen, - - -	do	-	do	-	-	-	-	-	-	-	-	-	-

FRANCE.

						yrs.							
Versailles, - - -	1848	Superior.	National Agronomic Inst.	-	-	-	9	9	-	-	-	-	3650
Grignon, - - -	1827	Intermediary.	Theoret. and Prac. Agr.	138	16 to 17	3	6	5	80	600	-	-	750
Grand Jouan, - -	1830	do	do	138	do	3	6	6	29	-	-	-	750
La Saulsaie, - -	1842	do	do	138	do	3	6	6	-	-	-	-	382
Saint Angeau, - -	1849	do	do	138	do	3	6	6	-	-	-	-	2247
Grand Jouan, - -	1830	Inferior.	Practical Agriculture.	-	16	4	-	-	29	-	-	-	-
Les Trois Croix, - -	1832	do	do	55	16	-	-	-	30	-	-	-	-
Salgourde, - - -	1839	do	do	-	16	3	-	-	30	-	992 an.	-	750
Montaurone, (La) - -	1839	do	do	110	16	4	-	-	21	-	1114 an.	-	420
Petit Rochefort, - -	1842	do	do	92	16	2 to 3	-	-	-	-	920 an.	-	325
Poussery, - - -	1843	do	do	55	16	3	-	-	33	-	773	-	950
Le Camp, - - -	1844	do	do	37	16	3	-	-	23	-	-	-	187
Le Corée, - - -	1845	do	do	-	16	4	-	-	15	-	-	-	-
Besplas, - - -	1847	do	do	-	16	3	-	-	23	-	360	-	270
Charmoise, (La) - -	do	do	do	-	16	-	-	-	24	-	-	-	-
Mesnil St. Firman, - -	do	do	do	-	16	4	-	-	12	-	368	-	692
Blanchampagne, - -	do	do	do	-	16	-	-	-	23	-	-	-	-
Calcomier, - - -	do	do	do	-	16	3	-	-	22	-	-	-	492
Trevarez, - - -	do	do	do	-	16	-	-	-	20	-	-	-	-
Bazin, - - -	do	do	do	-	16	-	-	-	15	-	368	-	225
Villechaise, - - -	do	do	do	-	16	3	-	-	17	-	220 an.	-	750
Chavaignac, - - -	do	do	do	-	16	3	-	-	22	-	-	-	970
Orme-du-Pont, - -	1848	do	do	-	16	-	-	-	13	-	756	-	215

FRANCE—CONTINUED.

Locality of the School.	Year when established.	Rank of the School.	Character of the Instruction.	Price of Tuition and Board.	Age of Pupils.	Length of the Course.	Number of Professors.	Number of the Courses.	No. of Students. Now in school.	No. of Students. From begin'g.	Paid by Government.	Scholarships.	Acres in the Farm.
Montbemeaune, -	1848	Inferior.	Practical Agriculture.	-	16	3yrs.	-	-	15	-	-	-	475
La Chauviniere, -	do	do	do	-	16	3	-	-	20	-	368	-	190
L'Espinasse, - -	do	do	do	-	16	3	-	-	17	-	-	-	550
Lajarrige, - -	do	do	do	-	16	3	-	-	15	-	-	-	-
L'Arena, - -	do	do	do	-	16	3	-	-	5	-	-	-	180
Visens, - -	1849	do	do	-	16	3	-	-	11	-	101	-	600
Petit-Metray, -	do	do	do	-	16	3	-	-	9	-	-	-	200
Berthauld, - -	do	do	do	-	16	3	-	-	9	-	368	-	182
Le Quesnay, - -	do	do	do	-	16	3	-	-	13	-	276	-	-
Carlan, - -	do	do	do	-	16	3	-	-	8	-	-	-	255
Le Courant, - -	do	do	do	-	16	3	-	-	12	-	572	-	125
Pergaux, - -	do	do	do	-	16	3	-	-	11	-	-	-	250
Guisancourt, - -	do	do	do	-	16	3	-	-	11	-	-	-	375
Petit-Chene, - -	do	do	do	-	16	3	-	-	11	-	-	-	417
Puilboreau, - -	do	do	do	-	16	3	-	-	11	-	-	-	375
Marolles, - -	do	do	do	-	16	3	-	-	8	-	-	-	1000
Germainville, -	do	do	do	-	16	3	-	-	10	-	529	-	250
Patris, - -	do	do	do	-	16	3	-	-	11	-	-	-	225
Mas-le-Comte,	do	do	do	-	16	3	-	-	12	-	-	-	225
Varincour, -	do	do	do	-	16	3	-	-	16	-	-	-	255
Templeuve, - -	do	do	do	-	16	3	-	-	11	-	2944	-	270
Les Bruyeres, -	do	do	do	-	16	3	-	-	9	-	-	-	650
La Roche, - -	do	do	do	-	16	3	-	-	7	-	-	-	332
Lamathe, - -	do	do	do	-	16	3	-	-	13	-	631	-	540

Trecesson, - -	1849	Inferior.	Practical Agriculture.	-	16	yrs.	-	-	9	-	-	-	352
Ollwiller, - -	do	do	do	-	16	3	-	-	11	-	-	-	265
Sainte-Crox, - -	do	do	do	-	16	3	-	-	9	-	-	-	312
Les Celestins, -	do	do	do	-	16	3	-	-	11	-	-	-	175
Lachaise, - -	do	do	do	-	16	3	-	-	10	-	-	-	337
Les Monts, - -	do	do	do	-	16	3	-	-	22	-	-	-	412
Saint Robert, -	do	do	do	-	16	3	-	-	11	-	276 an.	-	100
Saint Gildas, - -	do	do	do	-	16	3	-	-	11	-	-	-	270
Montat, - -	do	do	do	-	16	3	-	-	11	-	-	-	300
Montceaux, - -	do	do	do	-	16	3	-	-	19	-	276	-	512
Lahayereux, - -	do	do	do	-	16	3	-	-	10	-	-	-	725
Pont-de-Veyle, -	do	do	do	-	16	3	-	-	9	-	-	-	125
La Vignette, - -	do	do	do	-	16	3	-	-	6	-	-	-	388
Royat, - -	do	do	do	-	16	3	-	-	10	-	-	-	210
Villeneuve, - -	do	do	do	-	16	3	-	-	22	-	-	-	-
Beyrie, - -	do	do	do	-	16	3	-	-	11	-	-	-	300
Paillerois, - -	do	do	do	-	16	3	-	-	5	-	-	-	750
Mably, - -	do	do	do	-	16	3	-	-	11	-	-	-	2000
Castelbajac, - -	do	do	do	-	16	3	-	-	-	-	55	-	350
Salgues, - -	do	do	do	-	16	3	-	-	11	-	1104	-	875
Tolon, - -	do	do	do	-	16	3	-	-	11	-	-	-	245
Nolhae, - -	do	do	do	-	16	3	-	-	10	-	72	-	177
Souliard, - -	do	do	do	-	16	3	-	-	11	-	-	-	-
Mandoul, - -	do	do	do	-	16	3	-	-	-	-	-	-	272
Les Plaines, - -	do	do	do	-	16	3	-	-	-	-	-	-	-
Saint Just, - -	do	do	do	-	16	3	-	-	-	-	268	-	202
Castillaouenan, -	do	do	do	-	16	3	-	-	15	-	-	-	255
L'Hopital, - -	do	do	do	-	16	3	-	-	-	-	-	-	500

ITALY.

Pisa, - -	1842	Superior.	Attached to the Univers'y.	-	-	3 yrs.	-	-	-	-	-	-	-
Meleto, - -	1834	Inferior.	-	-	-	-	-	-	-	-	-	-	-

BELGIUM.

Locality of the School.	Year when established.	Rank of the School.	Character of the Instruction.	Price of Tuition and Board.	Age of Pupils.	Length of the Course.	Number of Professors.	Number of the Courses.	Number of Students. Now in school.	Number of Students. From begin'g.	Paid by Government.	Scholarships.	Acres in the Farm.
Tirlemont, -	1849	Intermediary.	Attached to a Communal College.	83—board.	15	3 yrs.	5	-	-	-	-	-	-
Vilvorde, - -	do	Special.	School of Horticulture.	Board free.	-	3	-	-	-	-	-	-	-
Oudenbourg, -	do	Inferior.	School of Practical Agricul.	55—board.	14	3	3	-	-	-	37 each pupil.	15	-
Chimay, - -	do	Intermediary.	Attached to College.	64—board.	-	3	5	-	-	-	-	6	-
Verviers, - -	do	do	do	-	15	3	2	-	-	-	-	-	-
Bastogne, - -	do	Inferior.	School of Practical Agricul.	-	16	3	2	-	-	-	46 each pupil.	-	-
Leuze, - -	do	Intermediary.	School of Agriculture.	83—board.	15	3	3	-	-	-	-	-	-
Lierre, - -	do	do	Attached to Communal Col.	64—board.	14	3	5	-	-	-	-	7	-
Attert, - - -	do	do	do	66—board.	-	3	3	-	-	-	-	10	-

PRUSSIA.

Locality of the School.	Year when established.	Rank of the School.	Character of the Instruction.	Price of Tuition and Board.	Age of Pupils.	Length of the Course.	Number of Professors.	Number of the Courses.	Number of Students. Now in school.	Number of Students. From begin'g.	Paid by Government.	Scholarships.	Acres in the Farm.
Moeglin, - -	1806	Superior.	Agronomy, Theoretical and Practical.	241	-	-	4	7	20	511	-	-	2480
Potsdam, - -	1823	Special.	Horticulture.	-	-	-	6	9	25 to 30	-	-	-	-
Frankenfeld, -	1825	do	For training Shepherds.	-	-	1	-	-	10	-	-	-	-
Eldena, - -	1827	Superior.	Agronomy, connected with University of Greisewalde.	129	17	2	9	10	80	-	-	-	1125
Regenwalde, -	1842	do	do	221	-	2	4	10	-	-	-	-	1125
Proskau, - -	1847	do	Practical Agriculture.	70—tuition.	-	2	8	8	-	-	-	-	2312
Reisen, - -	1847	Intermediary.	do	69—tuition.	-	2	-	-	-	-	-	-	-

Ragnit, - -	1847	Intermediary.	Agricultural Arts.	-	-	-	-	-	-	-	-	-	-
Grameuz, - -	do	Special.	Prairies and Irrigation.	-	-	-	-	-	16 min.	-	-	-	-
Siegen, - -	do	do	do	-	-	-	-	-	-	-	-	-	-
Poppelsdorf, -	1846	Superior.	Agronomy, Theoretical and Practical.	97—tuition.	-	2	9	10	47	-	-	-	70
Kolno, - -	-	Inferior.	Practical Agriculture.	-	-	-	-	-	12 max.	-	690	-	-
Near Strasburg, -	1845	do	do	-	-	-	-	-	-	-	-	-	-
Gliechou, -	1846	do	do	-	-	-	-	-	12 max.	-	-	-	-
Alach, - -	do	do	do	-	-	-	-	-	-	-	-	-	-
Reiffenstein, -	do	do	do	-	-	-	-	-	-	-	-	-	-
Bedersleben, -	do	do	do	-	-	-	-	-	-	-	-	-	-
Reisenrodt, -	do	do	do	-	-	-	-	-	-	-	-	-	-
Merchingen, -	do	do	do	-	-	-	-	-	-	-	-	-	-
Hasenfelde, -	do	do	do	-	-	-	2	-	15 min.	-	-	-	-
Schellin, - -	do	do	do	-	-	-	-	-	-	-	-	-	-
Grosskrebs, -	1847	do	do	-	-	-	-	-	12 min.	-	-	-	-
Lubtow, - -	-	do	do	-	-	-	-	-	-	-	-	-	-
Simmenau, -	1845	Special.	Culture of Flax.	-	-	-	-	-	245	-	-	-	-
Klopschen, -	1846	do	do	-	-	-	-	-	23	-	-	-	-
Nikolstadt, -	1847	do	do	Free.	-	-	-	-	-	-	-	-	-
Insterburg, -	1846	do	do	-	-	-	-	-	56	-	-	-	-
Troechtelborn, -	do	do	do	-	-	-	-	-	-	-	420	-	-
Zobter, - -	1845	do	do	-	-	-	-	-	-	-	-	-	-
Bidefield, - -	-	do	do	-	-	-	-	-	-	-	-	-	-
Herfort, - -	-	do	do	-	-	-	-	-	-	-	-	-	-
Rustern, - -	-	do	do	-	-	-	-	-	-	-	-	-	-

AUSTRIA.

Locality of the School.	Year when established.	Rank of the School.	Character of the Instruction.	Price of Tuition and Board.	Age of Pupils.	Length of the Course.	Number of Professors.	Number of the Courses.	Number of Students. Now in school.	Number of Students. From begin'g.	Paid by Government	Scholarships.	Acres in the Farm.
Krummau, - -	1799	Superior.	Agriculture.	Free.	-	-	-	-	-	-	-	-	300000
Prague, - -	1803	do	Industrial.	-	-	-	7	8	-	-	-	-	-
Grætz,(Johanneum)	1809	do	Agronomy.	-	-	-	9	11	-	-	-	10	-
Trieste, - -	do	Inferior.	Agriculture.	-	-	-	-	-	-	-	-	-	-
Lemberg, (University)	do	-	Professorship of Agriculture in University.	-	-	-	-	-	-	-	-	-	-
Trutsch, - -	do	Inferior.	Practical Agriculture.	-	-	-	-	-	-	-	-	-	-
Kœnigsâel, -	1835	do	Agriculture Industrial.	-	-	1 yr.	-	-	60	-	-	-	-
Geyesberg, -	1845	Special.	Culture, &c., of flax.	-	-	-	-	-	-	-	-	-	-
Twenty schools in different places.	1847 1848	do	do	-	-	-	-	-	-	Many thousands.	-	-	-
In Silesia, - -	do	do	do	-	-	-	-	-	-	-	-	-	-
In Moravia, -	do	-	-	-	-	-	-	-	-	-	-	-	-
In Galicia, - -	do	-	Professorship in University of Lemberg.	-	-	-	-	-	-	-	-	-	-
Cracow, - -	1848	Superior.	Agronomy.	-	-	-	-	-	-	-	600 ann.	-	-
Several schools for flax in Austrian Poland,	-	Special.	Culture of flax.	-	-	-	-	-	-	-	-	-	-

WURTEMBERG.

Hohenheim, -	1818	Superior.	Agriculture and Forests.	24 to 120	-	-	8	40	120	1650 to 1889	-	-	825
Do. - -	do	Inferior.	Practical Agriculture.	-	17	-	-	-	25	-	-	-	-
Do. - -	1844	Special.	Horticulture.	Free.	17	-	-	-	-	-	-	-	-
Do. - -	1845	do	Meadows; Irrigation.	Paid for work.	-	-	-	-	-	-	-	-	-
Do. - -	do	do	Fruit Trees; Flax; Instruments.	do	-	-	-	-	15	-	-	-	-
Ellrangen, - -	1843	Intermediary.	Practical Agriculture.	Board free; Tuition 12	17	3	-	-	20	50	-	-	-
Ochsenhausen, -	do	do	do	do	17	3	-	-	20	50	-	-	-

BAVARIA.

Schlessheim, -	1822	Superior.	Theoretical and Practical.	60	16	2	11	11	-	-	-	-	6115
Do. - -	do	Inferior.	Practical Agriculture.	24	-	1	-	-	-	-	-	-	-
Lichtenhof, -	1833	Inferior and Special.	Agriculture; Shepherds; Laborers.	-	-	3	-	12	26	-	-	-	-
Thirty-two schools in different places.	do	Inferior.	-	-	-	-	-	-	-	-	-	-	-

SAXONY.

Tharand, - -	1816	Superior.	Agriculture; Forests.	20 to 30	18 to 30	2½	9	10	80	1100	-	-	7437
Groshennersdorf,	1838	Special.	Small Culture, (orphans.)	-	8	6	-	-	80	150	-	-	67
Dresden, - -	1847	Intermediary.	Agriculture; Forests.	175	-	1 or 2	-	7	12	-	-	-	-
Broesa, - -	do	do	Agriculture.	140	-	1	7	8	12	19	-	-	-
Schoenfeld, -	1849	do	Practical Agriculture.	52	17	2	6	7	-	-	-	For 1-5 of the pupils.	

BRUNSWICK.

Locality of the School.	Year when established.	Rank of the School.	Character of the Instruction.	Price of Tuition and Board.	Age of Pupils.	Length of the Course.	Number of Professors.	Number of the Courses.	No. of Students. Now in school.	No. of Students. From begin'g.	Paid by Government.	Scholarships.	Acres in the Farm.
Brunswick, -	1835	Superior.	Agronomy; Technology.	-	16	-	13	13	100	-	-	$420 devote	d to schol
Schepenstedt, -	1844	Inferior.	Practical Agriculture.	56	-	2	-	3	30	-	-	-	-
MECHLENBURG SCHWERIN.													
Carlshof, - -	1844	Intermediary.	Theoretical Agriculture.	-	-	-	-	10	-	-	-	-	-
SCHLESWIG HOLSTEIN.													
Jevenstaed, -	1842	Inferior.	Practical Agriculture.	-	-	-	-	-	40	-	-	-	-
Rendsbourg, -	do	do	Agriculture—Practical.	-	-	-	-	-	40	-	-	-	-
Rodding, - -	1844	Mixed.	Agriculture.	-	-	-	-	-	20	-	-	-	-
Toestrup, - -	1845	Intermediary.	Agriculture—Horticult.	-	-	-	-	9	20	-	-	-	-
PRINCIPALITY OF ANHALT.													
Cochsted, 2 Div.,	1837 1844	Intermediary.	Agriculture—Irrigation.	200	-	1	-	13	-	-	-	-	-
Coswig, (Ascania)	1847	Superior.	Agricultural Arts.	-	-	1	10	-	-	-	-	-	-

GRAND DUCHY OF HESSE.

Darmstadt, - Arnsbourg, -	1831 1848	Intermediary. do	Agricult'l and Industrial. Practical Agriculture.	- 240	- -	- -	- -	- -	- 12	- -	- -	- -	- -

GRAND DUCHY OF WEIMAR.

Jena, - -	1826	Superior, connected with University of Jena.	Agronomy—Forests.	-	-	2	-	20	-	-	17 per pup'l	-	-

DUCHY OF NASSAU.

Geisberg, - -	1835	Superior.	Theoretical Agriculture.	Tuition 60	16	2 winters.	-	10	50	-	-	6	75

ELECTORATE OF HESSE.

Beberbek, - -	1846	Superior.	Agriculture—Forests.	119	16to20	2	-	8	16to20	-	-	-	-

GRAND DUCHY OF BADEN.

Hochbourg, -	1848	Mixed.	Agriculture of the country.	70 24 to 46 Tuition.	16	3	-	10	15	-	-	-	-

SAXE MEININGEN.

Locality of the School.	Year when established.	Rank of the School.	Character of the Instruction.	Price of Tuition and Board.	Age of Pupils.	Length of the Course.	Number of Professors.	Number of the Courses.	Number of Students. Now in school.	Number of Students. From begin'g.	Paid by Government.	Scholarships.	Acres in the Farm.
Frauenbreitungen,	1847	Elementary.	Practical Agriculture.	-	14	-	-	-	-	-	-	-	-

RUSSIA.

Locality of the School.	Year when established.	Rank of the School.	Character of the Instruction.	Price of Tuition and Board.	Age of Pupils.	Length of the Course.	Number of Professors.	Number of the Courses.	Number of Students. Now in school.	Number of Students. From begin'g.	Paid by Government.	Scholarships.	Acres in the Farm.
Marjino, - -	1824	Intermediary.	Agriculture; Mines; Forests.	450 (600 roubles.)	-	-	20	-	125	-	-	-	-
———, - -	1828	Inferior.	For the rearing of Bees.	-	-	-	-	-	253	-	-	-	-
St. Petersburg, -	1833	Superior.	Agronomy.	-	15	5	-	-	250	-	-	-	330
		Superior.	Agronomy.	84	15	3			85				1000
Gorigoretz, -	1836	Intermediary.	Mixed.	65	to	3	16	-	35	-	-	-	1000
		Inferior.	Practical.	38	20	3			200				1000
Kasan, - -	1840	Intermediary.	do	-	-	-	-	-	150	-	-	-	1935
Sympherpol, -	1843	-	For Silk Worms. (Magnanery)	-	-	-	-	-	150	-	-	-	-
Marjino, - -	do	Intermediary.	Practical.	-	-	-	-	-	-	-	-	-	-
Moscow, - -	do	Superior.	Agronomy.	80 to 100	17	-	-	20	-	-	-	-	660
Studener, - -	do	Intermediary.	Horticulture.	-	-	5	-	-	24	-	-	-	-
Six Farm Schools in different places,	do	Inferior.	Practical.	-	-	4	-	-	100 to 200	-	-	-	375 to 2000
Fifty Model Farms in different places,	do	do	do	-	-	-	-	-	-	-	-	-	375 to 2000
School for Flax, -	do	do	do	-	-	-	-	25	-	-	-	-	-

CONCLUSION BY THE COMMISSIONERS.

With these views and statements, the commissioners having had the various subjects committed to them under consideration, and feeling that the great object of agricultural education is one of vital importance, not only to the farmers of Massachusetts but to the prosperity of the Commonwealth, beg leave to submit the subjoined recommendations, and earnestly to commend the whole subject to the favorable attention of the Legislature.

But, inasmuch as it has been the custom of the State to disburse her funds for educational and charitable purposes, so as to encourage and enlist private munificence in conjunction with legislative aid, your commissioners offer their recommendations on the following conditions:

1. To carry out the foregoing views, the commissioners recommend the appropriation by the Legislature of twenty thousand dollars, for the purpose of establishing a Central Agricultural College, with a Model and Experimental Farm; said institution to be open to all classes of the Commonwealth, and in the government of which the State shall be interested so far as may be deemed expedient; provided, however, that this sum shall not be drawn for until an equal amount shall have been raised by private donation or legacy, and deposited in the treasury of the Commonwealth, to constitute a common fund for this object.

2. Whenever any incorporated Academy, not exceeding one such institution in each county, shall raise a fund of two thousand dollars or more, towards establishing and supporting a department of agricultural instruction, with lands suitable for experiments, it shall be entitled to draw from the treasury of the Commonwealth, the sum of two hundred dollars annually; provided, however, that whenever the number of scholars receiving instruction in agriculture shall be less than ten, the aid of the State shall be withheld.

3. The undersigned recommend the establishment of a State Department of Agriculture, to consist of a board of commission-

ers and a secretary, whom they shall annually appoint, which board shall sustain a similar relation to agriculture and the schools connected with it, as the board and secretary of education do to primary schools. This board shall consist of one member from, and to be elected by, each of the incorporated agricultural societies now receiving the bounty of the State, which board shall have power to locate, organize and put in operation the College contemplated by the foregoing recommendations. The duties of the secretary shall be, under the direction of the board, to give lectures in the various parts of the Commonwealth, whenever it may be deemed expedient, on the science and practice of agriculture; to receive the returns of the incorporated agricultural societies, and make a digest of the same in the form of an annual report to the Legislature; to collect agricultural statistics and information in the various departments of this science; to correspond with local societies in this and other lands; to visit individually, in connection with the board, the exhibitions of the various county societies; and to promote, by such other measures as the board may devise, this most important branch of human industry.

4. The commissioners further recommend, that inasmuch as the aid now rendered by the Commonwealth to agricultural societies was granted with reference to a very different state of things, when the population was smaller, agricultural products fewer, and all industrial pursuits were sources of much less revenue to the State; and in view of the increase of population, productions and revenue, an *additional* grant of one hundred dollars annually be made on every thousand dollars of the permanent fund of the several agricultural societies, which are or may be entitled by the present laws to the bounty of the State. Provided, however, that this sum shall not exceed nine hundred dollars to any society per annum.

5. That a premium, of such an amount as the Legislature may deem judicious, be offered for the best Elementary Treatise on Agriculture, suitable for common schools; said premium to be awarded by the Board of Agriculture, if such be created, or, if not, by a committee to be appointed by the Governor and council.

6. That after the common school fund shall have reached the sum of one million of dollars, and the Western Railroad sinking fund shall have been adequately secured, all proceeds of lands belonging to the Commonwealth in the state of Maine, and of the claims of Massachusetts on the federal government, shall be reserved to form a fund, the income of which shall be appropriated, at the discretion of the Legislature, for the encouragement or support of institutions for instruction in agriculture, for charitable purposes, and for education.

In conclusion, the commissioners have not deemed it expedient to report in favor of aid, at this time, to the other subjects committed to them for consideration, or to submit any more specific plans than those embraced in these general recommendations, or such as may be drawn from the substance of this report, leaving for the wisdom of the Legislature, or any Board of Commissioners which may hereafter be appointed, more perfectly to develop plans and modes of action.

Signed,

MARSHALL P. WILDER,
EDWARD HITCHCOCK,
SAMUEL A. ELIOT,
THOMAS E. PAYSON,
ELI WARREN,

Commissioners.

www.ingramcontent.com/pod-product-compliance
Lightning Source LLC
LaVergne TN
LVHW011119110826
845150LV00008B/2184
9781425506919